'All dogs have doggy genes, and cannot
undertake to be anything but dogs.
But each dog can be a dog in his own way.'

Dr Eric Berne

Acknowledgements

The authors are indebted to all those individuals who allowed themselves to be questionnaired, evaluated and counselled in the cause of developing the theories applied in this book; to Carl Jung for being interested in people's development; and, above all, to our wives for having to endure so many pompous lectures about why they behaved in that particular way at that particular time.

Individual
Excellence

Improving Personal Effectiveness at Work

Ralph Lewis and Phil Lowe

**KOGAN
PAGE**

First published in 1992. Reprinted 1996.

Kogan Page Limited
120 Pentonville Road
London N1 9JN

British Library Cataloguing in Publication Data

A CIP record for this book is available from the British Library.

ISBN 0-7494-0705-0

Typeset by DP Photosetting, Aylesbury, Bucks
Printed and bound in Great Britain by
Biddles Ltd, Guildford and King's Lynn

Contents

Introduction

We'd like to get one thing clear from the beginning: this book will *not* change your life. There are already a myriad ways of doing just that – most of which are regarded by the objective observer as being methods not of change, but of 'copping out'. Actions as distinct as cosmetic surgery and emigration both tend to involve ditching the bits of your world you dislike in the hope that what is left will somehow bring you to your personal El Dorado.

The constructive methods of achieving change – psychotherapy, education, religion – change not one's life but one's attitude to it. They give you extra tools with which to shape your ends. In our own work as management development practitioners, we use the twin tools of subject knowledge and self-knowledge to bring to managers a broader and deeper perspective on their working lives. We hope that this book will do the same for its readers.

People who talk openly of 'living in someone else's body' are usually regarded by the wider public as suitable cases for treatment. Yet every September, universities throughout the country welcome undergraduates coming up to study law because their fathers are magistrates, or to study medicine because their parents are GPs. The fact that these undergraduates would rather be studying ancient Iranian, or complex stresses in plastic mechanics, is irrelevant: their own personal preferences have been subordinated to the wishes of their environment – in this case, their families.

9

It can be argued, of course, that people in the above position cannot realistically expect to do everything they want to in this world. They may regard incurring their parents' wrath as a greater burden to carry than that of studying a course they do not enjoy. They may accept that their ability to support a family in the future may be more enhanced by a career in medicine than by operating a tower crane or pearl diving in the Philippines. This point of view is perfectly valid: we all have a responsibility to our outer world, even if it is only to turn up to work on time. There is a psychological contract which must be made between you as an individual and the world around you, for the mutual benefit of both parties; but such a contract cannot be made if you do not know exactly what it is that you want to gain from it. Most of us are fairly certain what it is we *don't* want, but find it hard to penetrate our own inner world any further than that.

Many of us – possibly the majority – stop at this point. We know we are dissatisfied with our life, but are not sure what to do next. What we often do is jettison all that we regard as being responsible for our unhappiness, in a knee-jerk reaction: it may not be as extreme as emigration or surgery; it may involve a hasty change of career or moving house to a neighbouring county. But, after all the upheaval is over, we find ourselves no happier; we may even grieve for what we have left behind or lost. Why?

Enter Carl Jung, one of the founding fathers of modern psychology, whose work forms the model on which this book is based. Jung said that we are all born with an innate preference for certain ways of living and working, just as we are born with an innate preference for writing with our left or right hand. If we choose to, and with a fair amount of effort, we can become ambidextrous; but if we are forced against our will to write with the hand we do not prefer (as many left-handed children have been, even comparatively recently) the result is unreadable, slow writing or, more serious, disturbances in hand–eye co-ordination.

The same is true of our innate personality preferences. Dissatisfaction or inefficiency in some aspect of our life is very often a result of having to use functions of our personality which are not those we would prefer. Dumping what makes us unhappy will only work if what we replace it with is matched to our true preferences – otherwise we are jumping from the frying-pan into the fire.

Because these preferences are innate, we cannot change them. What we *can* do is work at improving our ability to use our inferior functions, to use Jung's term for the less well developed area of our personality. The straight choice – find activities which match your preferences, or work on using the preferences which match your activities – is what we tackle in this book. The first step in any such enterprise is self-knowledge, which is why the questionnaire needs to be completed and scored before you turn your attention to the substance of the book. It is only by relating the areas the book covers to your own preferences that any realistic picture can be drawn.

This book, then, is unlikely to change your life. What we hope it may do is to help you to be happier with the one you're stuck with. At the end of the day, that has always seemed to us to be the preferable option.

1. Setting the agenda

A background to type theory

For as long as human beings have lived and worked together, they have been unable to resist the temptation to classify each other. The enduring nature of many classifications of personality 'type' – horoscopes being perhaps the obvious example – testifies to our need as individuals to feel that our day-to-day behaviour can be explained (some would say excused) by virtue of our falling into a particular category. The apparently 'disposable' nature of many modern horoscopes does not lessen the essential truth inherent in them. People *can* be classified. The important thing is to be clear about the point of classifying people.

When psychoanalysis became popular during the reign – or tyranny, depending on your allegiance – of Sigmund Freud as the king of psychological theory, the possibilities for classification widened. However, Freud's hypothesis that personality is governed by what happens to us as infants would tend to suggest an infinite variety of personality types: he denies us the possibility of predestiny so beloved of horoscope readers.

One of Freud's disciples, Carl Jung, took issue with Freud on this very point. He refused to see a person's whole way of life as being a way of sublimating various unsavoury childhood sexual experiences, asserting instead that a human being's mission in life is to develop and grow as a person. More importantly for our purposes, Jung proposed the theory that we are born with innate 'preferences' for the way we perceive (take in information) and judge (use the information to make decisions). These two simple processes are the mainstay of our day-to-day life, and Jung argued that the very clear differences in the ways individuals approach each process result in clear and identifiable personality types. Jung's model identifies eight types, to which we will return in the next chapter.

Many modern personality tests are derived from Jung's theory. By and large, they tend to preserve Jung's essential belief that the exercise of our innate preferences leads to a definite and identifiable 'type', one of a set of only eight (or sixteen, or what you will). The authors have taken a small but significant departure: we have chosen to encourage the readers of this book to look at the *balance* of different preferences within themselves, rather than to take a 'first-past-the-post' approach to classification. We are following the aspect of Jung's theory which suggests that we each use *all* of the functions he identified, with differing levels of skill.

Further explanation is necessary; but it is perhaps better to stop on the threshold of overload and leave you for a few minutes to ponder the questionnaire which follows. Once you have a set of figures in front of you at the end of the scoring process, we shall return more specifically to Jung's theory, and its implications for you *as an individual* – not a type.

Questionnaire

Instructions
For each of the following questions give 4 to the option that is closest to what you think, feel or would do. Give 3 to the next closest option, 2 to the next and finally 1 to the fourth closest option. There are no 'right' answers so just go by your own views. If none of the options says exactly what you want, choose the option which is closest.

Example

You are best at:

- ☐1 a taking action to deal with practical problems
- ☐ b motivating and influencing people
- ☐ c understanding people in depth
- ☐2 d analysing data
- ☐ e organising logical steps of action
- ☐3 f seeing possibilities in situations
- ☐ g doing factual detailed work
- ☐4 h seeing links between different areas

1. The way in which you learn best is by:

- ☐ a testing ideas or facts in reality
- ☐ b doing things practically
- ☐4 c feeling strongly about something
- ☐3 d seeing links and patterns in your mind's eye
- ☐1 e thinking through logical sequences
- ☐2 f being aware of different approaches
- ☐ g observing practical situations
- ☐ h discussing things with others

2. The subjects you most enjoy learning about are:

- ☐ a practical and down-to-earth
- ☐3 b to do with people and their actions
- ☐4 c to do with your own and others' values and feelings
- ☐ d in-depth and logical
- ☐ e those which apply principles in action
- ☐1 f creative and innovative
- ☐ g detailed and factual
- ☐2 h new insights

3. Your ideal work would involve:

- ☐3 a working with people and guiding them
- ☐ b working with technical equipment
- ☐ c analysing facts or figures to present logical solutions

[2] d working independently and creatively to develop new ideas

[ ] e problem-solving to improve practical situations

[4] f working by yourself with work that makes you feel good

[1] g gathering and collecting facts and figures

[ ] h developing creative ideas through discussions with others

4. You are best at:

[ ] a taking action to deal with practical problems

[ ] b motivating and influencing people

[4] c understanding people in depth

[ ] d analysing data

[ ] e organising logical steps of action

[3] f seeing possibilities in situations

[1] g doing factual detailed work

[2] h seeing links between different areas

5. When you have done something well you like appreciation for:

[2] a your ideas

[1] b your flair

[ ] c your accuracy

[ ] d your logical working through

[3] e your uniqueness

[4] f yourself as a person

[ ] g your commitment

[ ] h your practical skills

6. The strengths you bring to an organisation are:

[ ] a setting up good logical control systems

[ ] b caring for the people in the organisation

[ ] c sorting out practical problems as they arise

[ ] d seeing opportunities for the organisation in the market

☐ e looking at the overall running and links within the organisation
☐ f analysing and controlling data and information
☐ g working directly with people
☐ h keeping an eye on details

7. You judge organisations to be good if they:

4 a look after individuals' needs and feelings
☐ b are efficient in the way things are done
3 c encourage individuality and development
☐ d have means of planned achievement and growth
2 e promote good morale and team spirit in employees
1 f are flexible enough to take advantage of opportunities
☐ g understand the importance of logical planned systems
☐ h look after essential details

8. You would not like to work in an organisation that:

4 a paid no attention to individual development
1 b had no room for flexible responses to market opportunities
3 c did not put the needs of its people first
2 d ignored its social responsibilities
☐ e was inefficient in its use of control systems
☐ f didn't take time to analyse and plan
☐ g left things undone and ignored facts and details
☐ h was impractical in the way it did things

9. If you were having problems at work with a colleague, you would:

4 a talk to them and try to change their feelings
☐ b concentrate on doing a good job and ignore them
☐ c try to think of the logical thing to do
☐ d look for opportunities to change the way you work together
☐ e imagine different ways of looking at the situation
3 f keep your feelings to yourself

[2] g present the facts to the other person

[1] h have a reasonable discussion about the most logical thing to do

10. When you have a problem to solve the part you enjoy the most is:

[2] a finding out all the facts of the matter

[4] b looking at different possibilities

[ ] c looking at the problem from an unusual angle

[ ] d analysing the information logically

[1] e seeing how people feel about the issue

[2] f taking practical action to solve the problem

[ ] g presenting a logical plan of action

[ ] h getting others motivated to work together

11. In problem-solving you are best at:

[ ] a thinking of different options

[3] b taking people's feelings into account

[1] c seeing different aspects of the problem

[ ] d analysing the issues in depth

[ ] e getting others to take action

[4] f getting the facts

[2] g finding practical solutions

[ ] h presenting a logical plan of action

12. The type of people you would enjoy working with the most are:

[ ] a enthusiastic and motivating

[ ] b analytical and logical

[ ] c practical and realistic

[ ] d creative and innovative

[4] e caring and receptive

[2] f competent and successful

[3] g calm and pragmatic

[1] h insightful and unconventional

13. You would describe your managerial style as focusing on:

    ☐ a setting up and running good task control systems
    ☐2 b working with people to get the best from them
    ☐ c analysing data to run work efficiently
    ☐1 d seeing what customers and others want from your area
    ☐3 e practical present-day needs of the situation
    ☐ f looking for new ways to improve future performance
    ☐ g being realistic and adapting things for performance
    ☐4 h individuality and expression of that need

14. The managerial strengths you have are:

    ☐2 a realistic and pragmatic
    ☐ b innovative and flexible
    ☐1 c insightful and individualistic
    ☐ d analytic and efficient
    ☐2 e caring and receptive
    ☐ f active and practical
    ☐ g organised and principled
    ☐4 h fair and people-orientated

15. Your main strengths in a team are that:

    ☐3 a you see new opportunities
    ☐4 b you understand how people feel
    ☐ c you motivate and enthuse people
    ☐ d you can see clearly how to get results
    ☐ e you can take immediate practical action
    ☐ f you analyse complex factors in depth
    ☐2 g you see things from new angles
    ☐1 h you see details and facts clearly

16. You enjoy working in a team if:

    ☐4 a you feel that people are happy in the team
    ☐3 b there is a clear practical purpose to the team
    ☐ c it finds new and creative ways of doing things
    ☐2 d the team is achieving something well
    ☐1 e the team has good links with people outside it

☐ f it makes the most of every opportunity
☐ g it considers things logically and in depth
☐ h it completes tasks efficiently and well

17. You communicate best by:

☑2☑ a being receptive to people's feelings
☑1☑ b painting a clear picture of your views
☐ c presenting clear logical reasoning
☑4☑ d expressing your own feelings to others
☐ e giving a clear outline of the facts
☐ f understanding complex logical reasoning
☑3☑ g being receptive to other people's values and viewpoints
☐ h understanding the detailed facts of a situation

18. Areas you most enjoy discussing are:

☑4☑ a people's values and feelings
☑2☑ b practical experiences
☐ c general unconventional views and ideals
☐ d operational principles or ideas
☑3☑ e people's behaviour and attitudes
☑1☑ f new views and ideas
☐ g complex ideas and principles
☐ h down-to-earth realistic things

19. If you were trying to influence someone, you would:

☐ a be realistic and give the basic facts
☑4☑ b talk about future opportunities and benefits in broad terms
☑3☑ c present future ideals in terms of the benefit to individuals
☐ d give an in-depth analysis of the key factors
☑2☑ e talk about the effect on people's views and attitudes
☐ f present concrete benefits and factual improvements
☐ g present an overall analysis with logical benefits
☑1☑ h concentrate on people benefits such as morale improvement

20. Time is:

    ☐ a a resource to be used fully
    ☒ b a personal and subjective measure
    ☒ c a convention which we can rise above
    ☐ d a logical limitation on our capacities
    ☐ e a measure of what happens to us in life
    ☐ f a chance to experience things
    ☒ g how long it takes us to do things
    ☒ h the interval between two events

21. You are good at the following aspects of time management:

    ☐ a being flexible and responding to crisis
    ☐ b concentrating on getting things done now
    ☐ c responding well to interruptions
    ☐ d looking forward to the future's possibilities
    ☐ e allowing people time for their needs
    ☐ f working intensely when motivated
    ☐ g planning and being strict about time allocation
    ☐ h organising and keeping to deadlines

22. The things that tire you the most are:

    ☐ a concentrating on detailed factual work
    ☐ b having to take direct practical action
    ☒ c dealing with the demands of other people
    ☐ d having to create a range of future courses of action
    ☒ e coping with your own difficult feelings
    ☒ f coping with what seems to be a meaningless situation
    ☒ g having to make decisions on complex logical issues
    ☐ h structuring activities in a logical way

23. Stress makes it difficult for you to:

    ☐ a relax physically
    ☐ b do practical tasks
    ☐ c deal with people
    ☐ d understand and cope with new ideas or situations
    ☐ e cope with your feelings of being down

☐ f be yourself
☐ g think clearly or analyse things
☐ h organise or plan things to do

24. In a stressful situation you would:

☐ a look forward to doing things differently in the future
☐ b try to change the way you feel
☐ c try to understand what the stress is telling you
☐ d analyse the causes of the stress
☐ e get help from other people
☐ f try to beat the stress by switching off
☐ g take direct action to combat the stress
☐ h schedule activities to beat stress

25. You are happiest when you are:

☐ a having a good time with lots of people
2 b able to use your imagination fully
☐ c thinking about ideas and facts
☐ d doing something physically active
☐ e achieving your vision of what you want
4 f feeling happy and at peace with the world
1 g discussing ideas and facts with others
3 h physically relaxing

26. The things that make you feel great are:

1 a the companionship of others
☐ b understanding how things or ideas fit together
☐ c being on top of things through planning and thought
☐ d enjoying physically pleasurable experiences
☐ e achieving practical results through your skills
4 f being clear how your life is meaningful
2 g having a variety of successful and enjoyable projects
3 h the closeness of one or two special people

27. When you are on holiday, you want:

☐ a to be physically active

☑ b to lie back and enjoy the surroundings
☐ c to have a friendly time meeting lots of people
☐ d to be happy and content with one or two close friends
☑ e to have time to contemplate life
☐ f to absorb the different cultures
☑ g to have time to think about things or ideas
☑ h to plan different activities, sightseeing etc.

Scoring

From your responses to the 27 questions just completed you will obtain three separate profiles: **preferences**; **abilities**; and **stress**. The significance of each profile will be discussed in the following chapters.

Obtaining your profiles is a painstaking experience which some will find more rewarding than others (when you have read this book you will understand why!).

On the following pages you will find three score sheets corresponding to the three profiles. Each category gives the number of one or more specific questions from the questionnaire. For each question, place the points you allocated in the boxes beside the appropriate letters. When you have finished, follow the instructions below each score sheet to construct the profile matrix.

Preferences

		ES	EN	ET	EF	IS	IN	IT	IF
Learning	(q.2)	a ☐	f ☐1	e ☐	b ☐3	g ☐	h ☐2	d ☐	c ☐4
Work	(q.3)	b ☐	h ☐	e ☐	a ☐3	g ☐1	d ☐2	c ☐	f ☐4
	(q.5)	h ☐	b ☐1	a ☐2	g ☐	c ☐	e ☐3	d ☐	f ☐4
Organisations	(q.7)	b ☐	f ☐1	d ☐	e ☐2	h ☐	c ☐3	g ☐	a ☐4
	(q.8)	h ☐	b ☐1	e ☐	d ☐2	g ☐	a ☐4	f ☐	c ☐3
Problems	(q.9)	g ☐	d ☐	h ☐	a ☐	b ☐	e ☐	c ☐	f ☐
	(q.10)	f ☐	b ☐	g ☐	h ☐	a ☐	c ☐	d ☐	e ☐
People	(q.12)	c ☐	d ☐	f ☐3	a ☐4	g ☐2	h ☐1	b ☐	e ☐
Managerial	(q.13)	e ☐3	f ☐	a ☐3	d ☐1	g ☐	h ☐4	c ☐	b ☐2
Teams	(q.16)	b ☐3	f ☐	d ☐2	e ☐1	h ☐	c ☐	g ☐	a ☐4
Communication	(q.18)	b ☐2	f ☐1	d ☐	e ☐3	h ☐	c ☐	g ☐	a ☐4
	(q.19)	f ☐	b ☐4	g ☐	h ☐1	a ☐	c ☐3	d ☐	e ☐2
Time	(q.20)	g ☐3	a ☐	h ☐2	e ☐	f ☐	c ☐1	d ☐	b ☐4
Leisure	(q.25)	d ☐	e ☐	g ☐1	a ☐	h ☐3	b ☐2	c ☐	f ☐4
	(q.26)	e ☐	g ☐2	c ☐	a ☐1	d ☐	f ☐4	b ☐	h ☐3
	(q.27)	a ☐	f ☐	h ☐4	c ☐	b ☐3	e ☐2	g ☐1	d ☐
Total		11	11	16	21	9	31	1	42

Divide each total by 1.6 to convert it to a percentage.

	ES	EN	ET	EF	IS	IN	IT	IF
Total (%)	☐	☐	☐	☐	☐	☐	☐	☐
	1	2	3	4	5	6	7	8

Now take each percentage figure and transfer it to the following matrix. The number beneath each column above corresponds to a numbered space on the matrix. When you have transferred all eight figures, total them horizontally and vertically in the boxes provided.

	Extroverted	Introverted		Total
Sensing	1	5	*Sensing*	[ ]
Intuition	2	6	*Intuition*	[ ]
Thinking	3	7	*Thinking*	[ ]
Feeling	4	8	*Feeling*	[ ]
Total	[ ]	[ ]		

Now follow the same procedure for the next score sheet.

Abilities

		ES	EN	ET	EF	IS	IN	IT	IF
Learning	(q.1)	b ☐	f ③	a ①	h ④	g ☐	d ②	e ☐	c ☐
Work	(q.4)	a ☐	f ③	e ☐	b ☐	g ①	h ②	d ☐	c ④
Organisations	(q.6)	c ②	d ☐	a ☐	g ③	h ☐	e ☐	f ⑤	b ②
Problems	(q.11)	g ①	a ☐	h ☐	e ②	f ☐	c ☐	d ④	b ③
Managerial	(q.14)	f ☐	b ☐	g ☐	h ④	a ③	c ①	d ☐	e ②
Teams	(q.15)	e ☐	a ③	d ☐	c ☐	h ①	g ②	f ☐	b ④
Communication	(q.17)	e ☐	b ①	c ☐	d ④	h ☐	g ③	f ☐	a ②
Time	(q.21)	b ☐	d ☐	h ☐	c ☐	a ☐	f ☐	g ☐	e ☐
	Total	1	10	1	(14)	5	10	4	(15)

Divide each total by 0.8 to convert it to a percentage.

	ES	EN	ET	EF	IS	IN	IT	IF
Total (%)	☐	☐	☐	☐	☐	☐	☐	☐
	1	2	3	4	5	6	7	8

Transfer the figures as before.

	Extroverted	*Introverted*		Total
Sensing	1	5	*Sensing*	[ ]
Intuition	2	6	*Intuition*	[ ]
Thinking	3	7	*Thinking*	[ ]
Feeling	4	8	*Feeling*	[ ]
Total	[ ]	[ ]		

Stress

		ES	EN	ET	EF	IS	IN	IT	IF
Cause	(q.22)	b ☐	d ☐	h ☐	c ☐	a ☐	f ☐	g ☐	e ☐
Effect	(q.23)	b ☐	d ☐	h ☐	c ☐	a ☐	f ☐	g ☐	e ☐
Do	(q.24)	g ☐	a ☐	h ☐	e ☐	f ☐	c ☐	d ☐	b ☐

Score your answers but do not total them. There is no stress profile as such, but we will be using your scores for these three questions in the chapter dealing with stress.

Plotting your profile

As you work through this book you may find it helpful to have a visual representation of your profile. Figure 1 shows a quadrant on which you can plot the percentage totals from the preferences and abilities score sheets. We suggest treating each heavy dot as worth 5. Count out from the centre on each 'spindle', mark a cross for the appropriate value, then join the crosses to give a visual 'map'. Use a solid line for preferences and a dotted line for abilities (see the completed example on page 34).

How to interpret your scores

We were originally going to call this section 'What your scores

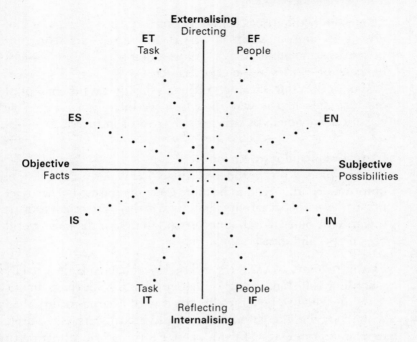

Figure 1

mean', but it would probably take most of the book to fulfil the requirements of such a title. This section aims to give the theoretical background which will enable you to begin to make sense of your questionnaire results. After explaining the theory in the next few pages, we will discuss how to 'read' the specific profile obtained from the questionnaire.

Jung's theory of psychological type, as we mentioned earlier, suggests that we all have innate preferences for certain ways of approaching life – just as we have an inborn preference for using our left or our right hand. As with our preferred hand, our preferred approach to life (or 'function' to use Jung's term) becomes with time more skilful, adaptable and *comfortable* to use.

Most people, when asked to say which hand they prefer using, and which hand they felt they used most ably, would name the same hand both times. With personality, however, it is not always so simple, which is why there are separate scores for preferences and abilities. We will be looking at the implications of these differences in the next chapter.

One of the central features of Jung's theory was the concept of our *attitudes*, or the ways in which we direct our energy. He classified individuals as either extroverted or introverted.

Extroversion and introversion

It was Jung who introduced these items into general use: unfortunately they tend to be more generally misused. They refer not to levels of sociability, but to whether our orientation is towards the outer world of people and things, or the inner world of concepts and ideas.

Extroverts move towards the external world and are attracted by what Jung called 'the object' – that which is outside them. In this sense they tend to be more concerned with external reality than with their own internal subjective reality. Extroverts have a drive to relate to this external world, and it is this active attraction to the world around them which leads many people to think of extroversion as synonymous with gregariousness. In fact it is perfectly possible to find shy extroverts. Extroverts are simply those who draw their energy from the world around them, and direct it back to the world. They can be thought of as under-stimulated: they move towards the world in order to be energised by it.

Introverts prefer their own internal world to the world around them. Their reality is a subjective, reflective reality. They can be thought of as being over-stimulated; they draw back from the world to avoid being overwhelmed by it. Introverts are not necessarily antisocial. They can have a strong need for other people, but it will tend to be for fewer, deeper relationships, whereas the extrovert might go for multiplicity. (In general, a drive for breadth is characteristic of extroverts, while introverts prefer depth.)

There are other basic differences which serve to clarify these preferences. Introverts tend to reflect before acting, whereas extroverts will tend to act first and think afterwards. Introverts will become drained by continued contact with others, and will need to seek solace in their own company in order to 'recharge their batteries'. Extroverts, on the other hand, are energised by contact with others, and if forced to work alone for long periods will very often have to seek company to energise themselves again. Extroverts are also more likely to wear their hearts on their sleeves, and to seem less complex than introverts: this is because introverts tend to turn their best sides inwards, as it were, to the sanctuary of their own inner world. Most extroverts would find it unthinkable not to apply their best characteristics to the world around them.

The four functions

Jung suggested that analysis of our daily activity can be reduced to the interaction of two distinct processes: **perceiving**, or taking in information; and **judging**, or making decisions based on the information received. There are two perceiving functions and two judging functions, and all four can be extroverted (directed outwards) or introverted (directed inwards).

The perceiving functions: sensing and intuition

Sensing is the practical function. As its name suggests, it relies on what can be perceived with our five senses. It is therefore concerned very much with concrete reality, and tends to disregard anything which cannot be seen, smelt, heard, touched or tasted.

- **Extroverted Sensing** is orientated towards external realities. People with a strong preference for extroverted sensing are often those who set out to experience everything that life has to offer – and to get the most from every one of those experiences. They will be practical, matter-of-fact, and orientated towards action.

- **Introverted Sensing** is also concerned with concrete experience, but in a more reflective way. Someone using this function would tend to take in an experience in order to understand its impact upon them as a person. This function often brings a strong awareness of one's own body and its internal physical processes. Depth of concentration, and an ability to focus on specific details, are also characteristic of introverted sensing.

Intuition concerns itself not with the facts of an experience, but with its implications, patterns and possibilities. Jung described this process as 'seeing round the object'. The intuitive looks at an object and asks not 'What is it?' but 'What might it be?' Intuition always looks to the possibilities of the future, whereas sensing tends to be rooted in the realities of the past.

- **Extroverted Intuition** looks to apply possibilities to the external world, whether this be of things or people. People with a strong preference for extroverted intuition tend to be good at seeing new opportunities in situations, at setting up new projects, and at spotting patterns or hidden meanings in data or events. This function often endows its owner with great enthusiasm for his or her vision of what might be. Stable or routine conditions tend to suffocate anyone with an emphasis on this function because of the lack of opportunity they provide.

- **Introverted Intuition** is equally concerned with possibilities, but without the requirement that they be related to any external application. This is the function of the poet, the seer and guru. Jung wrote that without this function there would be no prophets in the Old Testament. It is the most subjective and imaginative of all the functions, and is the function we use in pure brainstorming. People with a strong preference for introverted intuition will tend to be very unaware of their own physical concerns, being drawn instead by the depths of the human psyche, or by systems of belief which aim to give meaning to life.

The judging functions: thinking and feeling

Thinking is the rational function. It aims to bring order to things, people, ideas or concepts by measuring them against objective, non-personal data. Decisions made using the thinking function will invariably be logical – at least to the person making them.

- **Extroverted Thinking** seeks to organise the external world into logical patterns or systems. People with a strong preference for this function will often enjoy discussing ideas and theories with others in order to arrive at an agreed truth. They will also be good at analysing situations and events to find out what is happening and why. Extroverted thinking looks at the world in terms of cause and effect, in terms of principles and demonstrable truths.

- **Introverted Thinking** is equally logical, but the logic in question need not be universal. Someone with an emphasis on this function may be concerned only that things fit into their own pattern of logic or principles. Introverted thinking endows its owner with a capacity for comprehension, for absorbing events and fitting them into a conceptual model. It can also lead to a need to be clear in one's own mind about the principles involved before accepting an idea or task. Accordingly, along with the gift of objectivity, stubbornness may often be found in tow.

Feeling is not concerned with objectivity. Where thinking seeks order, feeling seeks *harmony*, and tends to favour decisions or solutions which cause the least disruption to people and their values. Where thinking asks if something is reasonable, feeling asks whether it is fair. Decisions are made by weighing up how much an issue matters, both to the decision maker, and to others involved.

- **Extroverted Feeling** leads to a concern for people and their values in general. Someone with this preference is likely to be highly sensitive to the feelings and tastes of others, as well as to the requirements of fashion and etiquette. The values which extroverted feeling emphasises are subjective, but not necessarily personal to the owner; they are just as often subjective values which are shared by society but which have no basis in logic (such as the correct things to say in a certain situation). This function will tend to lead to a desire to live in harmony with universal standards of fairness.

- **Introverted Feeling** relies on subjective values which are truly subjective. Someone with a strong emphasis on this function may be genuinely difficult to understand, and will find it hard to discuss his or her innermost values. Although such a person will probably care passionately about others and have a deep insight into their needs and feelings, they may still seem to an observer to be disinterested, as their concentration is generally on the intensity of what they personally feel inside. This function leads to a strong and clear sense of what matters to one as a person.

Your profile

When you answered each individual question, your scores (4, 3, 2 or 1) were automatically allocated to one of the eight functions – the eight columns on the score sheet. They are abbreviated as follows:

Extroverted Sensing	ES	Introverted Sensing	IS
Extroverted Intuition	EN	Introverted Intuition	IN
Extroverted Thinking	ET	Introverted Thinking	IT
Extroverted Feeling	EF	Introverted Feeling	IF

The totals you transferred into the final matrices give the percentage of the total scores allocated to each function. Below is an example of a completed preferences matrix.

	Extroverted		*Introverted*			*Total*
Sensing	1	1.4	5	6.4	*Sensing*	7.8
Intuition	2	16.1	6	27.1	*Intuition*	43.2
Thinking	3	7.1	7	6.8	*Thinking*	13.9
Feeling	4	8.9	8	26.2	*Feeling*	35.1
Total		33.5		66.5		

Percentage scores for the eight functions as given in the two central columns, so in this example introverted intuition has a score of 27.1 per cent.

Totals for extroverted and introverted functions combined are given in the right-hand column, so thinking in total has a score of 13.9 per cent.

Finally, an overall score for extroversion and introversion is given at the bottom. In this example the balance is extroverted 33.5 per cent, introverted 66.5 per cent.

By looking at this example profile it can be seen that this person has an overall preference for his/her inner world. The functions that he/she prefers using most are:

- Introverted Intuition
- Introverted Feeling
- Extroverted Intuition

The functions they least prefer using are:
- Extroverted Sensing
- Introverted Sensing
- Introverted Thinking
- Extroverted Thinking
- Extroverted Feeling

Using the functional descriptions given above, one can start to build up a picture of the way this person is likely to operate.

Figure 2 shows the same preferences plotted on to the quadrant. Once ability scores are taken into account, the picture becomes more intricate.

You will see that a large proportion of the profile lies in the internal–subjective quadrant. Using the quadrant profile can sometimes give more of an 'at-a-glance' guide to the basic thrust of a personality.

Reading the basic function descriptions given earlier in this chapter should ring some bells in terms of your own personality. Read again those descriptions which correspond to the functions you identified as your strongest preferences and see if they ring true. Then read through those you have identified as your weakest preferences. There may well be some points you disagree with, but most of it should make sense.

The rest of the book is designed to flesh out your profile even

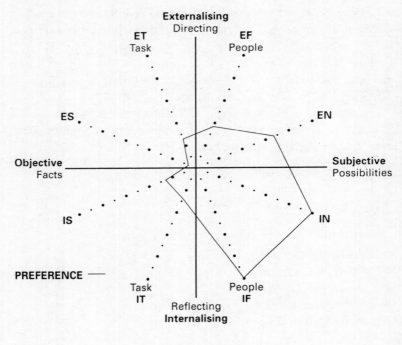

Figure 2

more, to enable you to reach a final conclusion about your own preferred operational style. Remember, this profile is based on your own answers, so the final judgement as to its accuracy rests with you.

2. Where am I?

Preference and ability: what's the difference?

Jung's theory of psychological type suggests, as we have seen, that we are born with an innate 'preference' for a particular way of operating, of approaching the world. The theory implies that, through consistent exercise of our preferred operational style, we become more adept at it, and therefore choose to use it even more consistently. Preference and ability, this theory implies, follow each other sequentially, and correlate almost completely with each other.

In the questionnaire you completed earlier, preference and ability are scored separately. Some readers' scores will be similar for both sets of results: their preference and ability profiles overlap. These people are the evidence one can cite in support of the 'type development' theory summarised in the paragraph above.

Many readers, however, will find that their scores for the two profiles are different, sometimes radically so. You may find that your preference scores are definitely orientated towards, say feeling and intuition (the subjective end of the quadrant model) but your ability scores lie more towards thinking and sensing (the objective end). Does this mean that you have got the questions 'wrong'? That you are very poor at self-awareness?

The answer to that question is complex and eventually inconclusive. No questionnaire is guaranteed to reveal the truth. It reports back what you have told it. Depending on your mood when you completed it, the environment in which you completed it, whether you were in a hurry, and any number of other reasons, your answers may not reflect the true you. Reassurance on that score will come from reading this book; the numerous descriptions of the effects of each preference in different situations should enable you to decide whether your reported profile is accurate for you – whether it 'feels right'.

Another point which should be made is that ability is very difficult to measure; it is even more difficult to rate in oneself. However, before you burn this book and dismiss the authors as tedious time-wasters, be assured that there is a benefit in attempting to assess your own ability. For a start, we often believe we are 'best' at those things which we seem to find ourselves doing most regularly. If the accountant from the office next door is always asking us to check her sums, we will tend to conclude that she must consider us to be good at maths (or, possibly, that she is in the wrong job). Assuming that we conclude the former, the next question to ask is: do we *enjoy* checking sums? Does it 'feel good'? In an ideal world, would we *choose* to check sums? The answers to these questions will take us towards establishing our *preference*. If we answer no, we will score differently on preference and ability when it comes to checking other people's maths.

If your scores are similar on preference and ability, your focus is likely to be on whether the work you do, and the environment in which you do it, are conducive to your preferred operational style – since you are also confident that that is where your abilities lie. If your scores on preference and ability are different, the first question to ask is: why? To answer that rather economical question, we must move rapidly into another convoluted argument. One possible reason for the difference is environmental. Let us consider the profile in Figure 3.

There is a clear tension in this person's profile between ability and preference. She is a researcher, who at the time she completed the questionnaire had been involved in one very detailed project spanning several months; it required her to work in isolation for long periods, sifting and organising data. It is not surprising, therefore, to find that according to this profile, she sees her

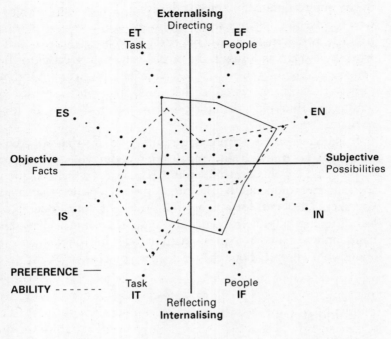

Figure 3

abilities as lying heavily on the objective (sensing–thinking) side – especially introverted sensing and introverted thinking. The area where her preference outstrips her perceived ability, on the other hand, is feeling, both extroverted (other people) and introverted (personal values). She must be in the wrong job, yes?

Actually, no. She is very happy in her work. What emerged during a counselling session based on her questionnaire results was that she had simply spent too long on one particular project. At the time she completed the questionnaire, she was desperate for an opportunity to broaden the scope of her job to include more client liaison: to get out and about, to get away from the data for a little while (you will see she has a preference for extroverted intuition, and so will tend to become bored if a project gets too samey). The end result was that she overcompensated when filling in the questionnaire, and emerged with a falsely high preference score on feeling: extroverted, because she felt starved of human contact; introverted, because she had got so bogged down in her work that she was losing sight of her values. Her rating of her abilities was simply the result of having been given an important project to work on which, to her mind, required heavy use of sensing and thinking – therefore, she concluded, she must be good at them.

It is because of situations and results like this that we advise you to avoid drawing any conclusions about your preferences before reading this section, and Chapter 3, which is about working environments. By the time you have reached that point you should feel comfortable with a profile of your preferences; any difference in preference and ability scores from the questionnaire should be used as a springboard to ask some tough questions about yourself and about the work that you currently do.

Learning styles

Without learning, it is said, there is no development. This is perfectly true, but very little comfort when trying to absorb Chapter 17 of *Principles of Accounting* on the 07:23 train from Matlock the week before a vital professional exam. The older we get, the more painfully aware we are of how easy it was to learn things as a child, when – of course – we did not appreciate our good fortune. But, as always, we have a choice. We can look out of the train window and ask 'Où sont les neiges d'antan?', or we can refer to our preference map on page 27 and compare it with the learning styles map in Figure 4.

Being clear about our learning preferences may mean we decide to throw *Principles of Accounting* out of the window and run off to Buxton, or realise that we would absorb it better, and learn more from it, via a group discussion for example.

For all eight functions, the following comments can be applied both to the preferred subjects of someone with that preference, and also the way in which he or she might learn best when dealing not only with a preferred subject but with learning in general.

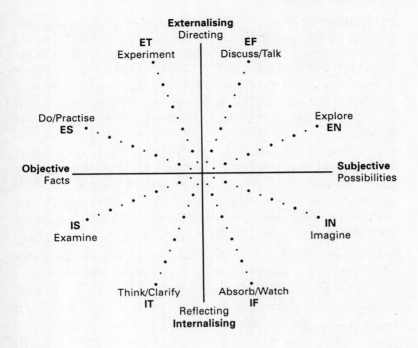

Figure 4 *Learning styles*

Reflective (introverted) learning styles
The reflective styles rely on the topic being studied producing a subjective resonance within the student. In other words, the learning experience is always related directly to the student, rather than to the outside world.

- Using the **introverted feeling** function to learn involves reflecting on what a subject means to you personally, in particular what you *feel* about it. If vocational subjects were literally vocational, they would be studied mainly by people with an introverted feeling preference. As it is, one studies a vocational subject usually because of some ultimate goal which may or may not be to do with one's own subjective feelings. If you find it hard to learn effectively when your own feelings are involved, your introverted feeling function may be under-developed.

- Using the **introverted intuition** function involves conceptualis-ing: making links and seeing patterns which are relevant for your personally. If you tend to use this function you may also prefer learning about subjects which involve imagination and insight. You may find you develop an instinctive understand-ing of some subjects long before developing full familiarity with them (don't forget, Jung himself said that if it were not for this function, there would be no prophets in the Old Testament). If you find it difficult to learn when called upon to use imagination, it may be worth aiming to develop your introverted intuition function.

- Learning using the **introverted thinking** function means setting up a system of logic to absorb the new subject – but remember that the focus of this logic is inward. This is not the scientist setting up an experiment: this is the theorist, working through a logical sequence of ideas and finding a niche for the new piece of learning. People with this preference often prefer learning about subjects which are in-depth and logical. If you have trouble thinking out things for yourself as part of the learning process, your introverted thinking function is likely to be underdeveloped.

- The **introverted sensing** function represents probably the most practical of the reflective learning styles – but still with a subjective focus. Using this function can involve observing practical situations and absorbing the lessons from them, homing in on the precise and detailed relevances in the situation being observed. Factual, detailed subjects are likely to appeal to someone with this preference, while someone with an underdeveloped introverted sensing function will probably find it difficult to learn when having to absorb detailed facts.

The directive (extroverted) learning styles
Here the focus is on external concerns and applications, on action rather than reaction.

- Someone using the **extroverted feeling** function in a learning situation will learn much better when working with other people. They will need to talk things out before they can understand them. Those who find it difficult to learn when working with a group may well have an underdeveloped extroverted feeling function. Those with this preference may also find it helpful to relate what they are learning to the shared values of their group or of their wider society. People-orientated subjects are likely to appeal to those with this preference.

- Developing new possibilities for the subject being studied involves the **extroverted intuition** function. If you are using this style the question to ask is 'What can I do with this?' or perhaps 'How might I do this differently or better?'. An extroverted intuition preference for learning often leads to an interest in subjects which have a creative, innovative character or application to the outside world. If you find it difficult to learn when playing around with possibilities, your EN function may be underdeveloped.

- Whereas the introverted thinking function allies with subjects which have a logical depth, the **extroverted thinking** function prefers broad principles which have a logical application. The workings and results of real systems, whether mechanical or organisational, may well hold an appeal for someone using this learning style. Those with this preference tend to learn best when given the ability to test ideas or facts in reality, to apply certain scientific principles to what is being learned. Someone in whom this function is underdeveloped might well find it difficult to learn when having to think about complex, real-life situations.

- Given that the **extroverted sensing** function is the most realistic and practical of the eight functions, it should be no surprise to discover that a preference for practical subjects is the hallmark of someone with this learning style. Those with this preference learn best by doing things themselves or observing practical examples. Abstract concepts are of no help. If you find it difficult to learn by getting straight into practical work, your extroverted sensing function may be at fault.

So the person trying to absorb *Principles of Accounting* by private reading alone will do very well if they are using a developed introverted thinking learning style. However, if that style is underdeveloped they will have great difficulty. Conversely, if a company sets up a workshop consisting of continual syndicate work and group discussion, those with an underdeveloped extroverted feeling function will find it difficult to learn anything.

Managers concerned with developing their staff, and educationalists of all types, have a tendency to think that the way they themselves learn is appropriate for everyone. This, of course, is not the case – unless their staff or pupils have the same psychological make-up as them. Carefully spelt out learning objectives, as required by those with thinking preferences, may be seen as straitjackets by those with intuitive preferences who want to learn in an exploratory, open way.

Freedom to learn in the way best suited to us means growth and development without pain. It means that learning can be fun – a concept that unfortunately seems to have been forgotten.

Values and motivation

The word 'motivation' comes from the same root as the verb 'to move'. Once money is removed from the picture, we all emerge as being motivated in our working lives by very different considerations. The basis of most individual motivation lies in the personal set of values that each of us has. We may not think of them in this way – it depends on individual preference. 'Values' as a word tends to have most appeal to those with a strong feeling preference. Someone with a preference for thinking might think of them as 'principles', intuition might prefer 'ideas', and sensing might talk of 'needs'. Whichever way you slice it, however, the fact remains that most of us are driven, somewhere along the line, by considerations other than money. Furthermore, we can break these down into individual 'career values' grouped according to the Jungian model with which we are working.

We have acknowledged that someone's profile can be distorted by environmental factors; personal values are, if anything, even more susceptible to influence by parents, organisations, education – anything that happens to you in your life can leave a mark which may be interpreted as your guiding principle. As a starting point, we have taken some basic motivational factors and grouped them according to the preference with which they are most compatible. Before we go through the values by function, we suggest you choose the eight values which are most important for you. Think of them in terms of an ideal job. Prioritise them from 1 (most important) to 8. Then check the extent to which they reflect your preferences by looking at the groupings. A brief description of each of the values is given below.

Adventure   Doing work that involves lots of risk-taking – as defined by you. It could be going up the Amazon or driving round the M25.

Aesthetics   Doing work that involves studying or appreciating the beauty of things or ideas.

Affiliation   Wanting to belong and work in a particular organisation – the organisation is more important than the task.

Artistic creativity   Working in any one of the artistic media – painting, writing, making music, etc.

Challenging problems   Problem-solving as a key component of work – applying logical principles to get solutions.

Change and variety   Doing work that varies from one day to the next – always different.

Community   Doing work that enables you to be part of the local community.

Competition   Pitting your wits and abilities against others.

Creative expression   Communicating views or ideas in a novel and different way.

Creativity   Using inspiration, flair, etc to develop products, ideas or services in ways not done before.

Excitement   Getting a 'buzz' from the work you do.

Exercise competence   Using your abilities to the full, but perhaps steering clear of areas in which you doubt your competence.

Fast pace   Handling lots of work quickly.

Friendship   Forming close personal relationships with people at work.

Helping others   Helping other people directly and on a personal basis.

Helping society   Doing something that you feel contributes to the fulfilment of society as a whole.

High earnings   Earning significantly more than the average for your job. If you select this, make sure you are thinking of it as a specific statement of what you want – not just 'it would be nice to be paid more'. Are you prepared to sacrifice other things to get more money?

Independence   Being able to do work in the way you want without having to follow rules or regulations or doing as others say.

Influencing others   Using persuasion and getting others to do things.

Intellectual status   Being regarded as an expert in a given field – someone to whom people come for advice.

Job tranquillity   Having a nice peaceful job that avoids pressure.

Knowledge   Dealing with data or information – processing and organising it logically.

Location   Living in a particular place, whether village, town, city or region.

Making decisions   Deciding courses of action or policies using your own judgement.

Moral fulfilment   Feeling that your work is contributing to your ideals and deeply felt personal values.

Physical challenge   Doing work that demands physical agility, strength, stamina or dexterity.

Planning and organising Scheduling and thinking through actions and plans in logical sequences.

Power and authority Enjoying being the boss and getting others to do things by virtue of your position.

Precision work Doing work that needs close and exact attention to detail.

Profit, gain Doing work that enables you to earn chunks of money rather than a regular salary.

Promotion Advancing quickly up the organisational hierarchy for work well done.

Public contact Spending a lot of time dealing with members of the public.

Recognition Getting public credit for work well done.

Security Being assured of keeping your job for as long as you like (a rare occurrence in reality these days!)

Stability Doing work that doesn't change from one day to the next – a reliable routine.

Status Wanting the respect of friends, family, people in general for the level or nature of your work.

Supervision Looking after others in terms of getting them to do specific work.

Time freedom Working the hours you want, starting and finishing when you want.

Work on frontier of knowledge Developing new ideas or hypotheses.

Work under pressure Overcoming obstacles and pressures such as dealing with awkward people, deadlines or sheer quantity of work.

In the following text the values have been sorted into their most likely functional 'homes'. For the purposes of this list, we have not broken the functions down into their extroverted and introverted forms; these attitudes tend not to affect individual values as such, more the way in which they are expressed, although a need to work in teams might sit better with an extroverted type, for example.

Some of the values are 'wild cards', which could be held by anyone regardless of preference. These have been listed separately.

- **Sensing**
  Adventure
  Aesthetics (also under **Intuition**)
  Excitement (also under **Intuition**)
  Fast pace
  Location
  Physical challenge
  Stability

- **Intuition**
  Aesthetics (also under **Sensing**)
  Artistic creativity
  Change and variety
  Creative expression
  Creativity
  Excitement (also under **Sensing**)
  Independence
  Time freedom
  Work on frontier of knowledge

- **Thinking**
  Challenging problems
  Competition
  Exercising competence
  Intellectual status
  Knowledge
  Making decisions (also under **Feeling**)
  Planning and organising

- **Feeling**
  Affiliation
  Community
  Friendship
  Helping others
  Helping society

Influencing others
Making decisions (also under **Thinking**)
Moral fulfilment
Public contact
Supervision

- **Wild cards** (may apply to any preference)
High earnings
Job tranquillity
Power and authority
Precision work
Profit, gain
Promotion
Recognition
Security
Status
Work under pressure

In more general terms, reading through the functions' definitions on pages 29–32 may trigger off associations with some of your own values: preferences and values are not essentially that far apart. We have introduced the topic of values and motivation early in the book because, paradoxically, it may be the last area with which the reader may come to terms. We suggest that as you read the rest of the book, one corner of your mind stays on this chapter, as many of the descriptions concerning work preferences will certainly be connected with the fundamental motivations which lead us all to love or loathe the work we do.

3. The working environment

Is your work self the same as your home self?

Carl Jung's theory of personality type allowed for the existence of the 'shadow' – that part of our personality which is outside conscious control and which tends to be a mirror image of our conscious self, ruled by those functions which are less developed. There is a school of thought which postulates that when we are at home, we live as our shadow selves – and thus create a 'home' self which is the opposite of our 'work' self. This is an attractive theory, but not one to which the authors subscribe. However, it is certainly true that the shadow is always looking for a way out, as it were – and the home is a much safer environment to parade our undeveloped aspects.

Horse Badorties, the amiable hippy hero of William Kotzwinkle's novel *The Fan Man*, introduces us to 'Uncle Skulky', the hideous old relative who dwells somewhere inside all of us. It is essential, Horse tells us, to allow Uncle Skulky out for some exercise every now and then, otherwise he will take you over. As we will see later when looking at stress, an unexercised and unfamiliar shadow self can pop out without warning, and isn't too fussy about where we happen to be at the time.

All things being equal, home is theoretically the place where we are most like ourselves; at home, we are in what personality enthusiasts like the authors are fond of referring to as a 'shoes off' state. This means, in effect, that when we are at home we are most likely to behave like our *preference* profile. The *abilities* profile is probably more influenced by the kind of work we do, and the things our employers convince us we are good or bad at.

The questions in the questionnaire were all work-related, and so your preference profile may be skewed slightly by what you desire specifically at work. The authors use professionally a longer version of the questionnaire which covers home and work, and in fact there tends to be little difference between the profiles – the only consistent difference being that work profiles tend to have a slight extroverted bias, reflecting perhaps the inherently social nature of most workplaces.

The main difference between a home preference and a work preference lies in the opportunity we find to apply our preferences. The 'in an ideal world' criterion attached to preference is more likely to reflect reality at home than at work – and many of the people reading this book would probably not be doing so if their preferences were as well served at work as at home.

Whereas our 'home self' – once children, spouses, friends or animals have been fed, hugged and sat in front of the TV – is concerned to a large extent with self-actualisation or gratification, our 'work self' exists as much for the gratification of others as of ourselves. In order to succeed at work we must fulfil our side of the psychological contract – that invisible and intangible, yet totally binding, document which index-links our fortunes to those of our employer; and the more dissatisfied workers tend to be those who do all the fulfilling but don't seem to get much back. These people's own needs and preferences become submerged beneath the brave face they put on things – get the job done, get the wage-slip, don't rock the boat – and, as many people know from experience, if you spend too long in a groove you become frightened even to look over the edges.

For the purposes of this chapter of the book, however, it is your work self that needs to be satisfied – whether or not it is a self you shed as you walk through the front door.

Working within an organisation

The purposes of organisations

The primary focus of any organisation can be thought of as

targeting one of four functions, which can be mapped on to a quadrant like the one shown in Figure 5.

The individual within

You will have noticed that the quadrant in Figure 5 is conveniently similar to the quadrant we use to map out the eight functions – but this similarity should be considered with caution. A strong preference for extroverted intuition does not guarantee that one will be happiest working in an organisation with an External–People focus. They are not incompatible, however: Figure 6 summarises the quality of each individual preference within a work setting.

- At work a preference for **extroverted sensing** focuses on the practical realities of situations, and in particular on the process rather than the results: practical action to deal with problems,

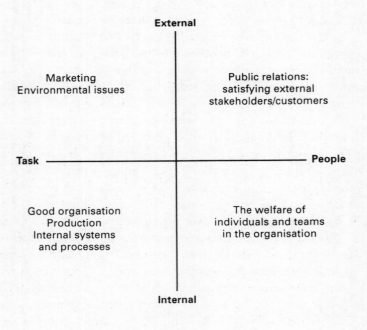

Figure 5

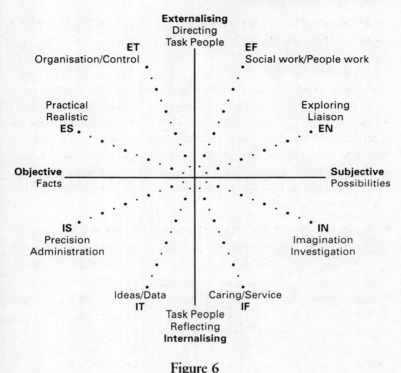

Figure 6

working with technical equipment, doing physical work – all of these require the use of extroverted sensing. Areas of work in which this function features largely include handicrafts, engineering, labouring, the military, and mechanics.

- **Extroverted intuition** takes the intuitive function's focus on possibilities and relationships and applies it to the external environment. There can be a number of target areas: developing creative ideas through discussion; looking at existing set-ups and seeing ways in which they might be enhanced; building advertising campaigns; and experimenting with new ideas. Professions which often require extroverted intuition include science, psychology, entertainment, design, architecture, and advertising.

- **Extroverted thinking** involves the instigation of logical systems of control, monitoring or analysis. When combined with extroverted sensing, it forms most people's image of the typical problem-solving process – step by step and logical. Divorced from sensing, it is more about precision in dealing with problems, about organising action into clear structures: work scheduling, for example, or coming up with reliable ways of measuring results. When used for decision-making at work, it tends to produce impersonal, logical decisions. Professional applications of this preference can be found in consultancy, computing, management, administration and marketing.

- **Extroverted feeling**, the most sociable and people-orientated function, tends to foster a desire to work with people, motivating and influencing them, promoting harmony between them, or organising them. Decision-making using extroverted feeling will tend to involve consultation with those concerned, and will weigh up the consequences of the decision for those affected by it. Professions which offer scope for using extroverted feeling include sales, secretarial work, nursing, teaching (especially pre-school and primary) and the theatre.

The introverted functions in a work environment tend to focus on the inner world of concepts and ideas, but are not necessarily antithetical to human contact. When an introverted preference is being applied to other people, the ideal is usually contact with one or two, preferably familiar, people at any one time.

- **Introverted sensing** usually implies a concentration on detail, but can also involve the gathering of facts and figures. Practical or physical work which does not rely on teamwork may also be attractive to people with a strong IS preference. This preference can be found in occupations as diverse as physiotherapy, word processing and farming.

- **Introverted intuition** is the imaginative function, and is concerned with the engendering and development of ideas – but ideas which do not necessarily have any application: it is up to the organisation to decide whether or not they are valid. As such, introverted intuition represents the starting point of the innovation process. Companies depicted in TV drama series often contain one maverick employee who seems to be paid a lot of money for wandering around acting strangely and occasionally having a brilliant idea. This is the person with a strong introverted intuition preference. This function can also be used to absorb the links and patterns inherent in data or situations, and finds expression in such professions as writing, research, lecturing and mathematics.

- **Introverted thinking** is as logical as its extroverted counterpart, but is more concerned with the absorption of logic than with its application. It therefore lends itself to independent analysis of data or situations, to the creation of conceptual models, and to the evaluation of ideas and plans within a logical framework. However, the logic applied by someone with this preference is not necessarily universal; it can be a very personal system of logic, and so workers with a strong IT preference may seem as abnormal as the introverted intuitive mentioned in the paragraph above. Professions which benefit from the use of introverted thinking include accounting, analysis, auditing and electrical engineering.

- **Introverted feeling** finds its application in more than one way. In terms of the need for self-actualisation, which frequently characterises people with a strong IF preference, it is often expressed in solitary work which does no more than make the person doing it 'feel good'. In terms of the capacity for empathy which it endows, this preference is often used in work which involves an insight into the feelings of individuals, or in one-to-one counselling situations. Apart from counselling, other occupations which draw on this preference include religious and social work.

The organisational environment

Ignoring those specific careers mentioned on the previous pages as pointers, it is clear that most organisations will need individuals representing all eight preferences. However, many organisations will have an environment, or character, more specifically identifiable as related to one or more of the functions.

It has to be said that one cannot always expect total compatibility with the organisation of which one is a part. The 'traditional' British organisation tends to score high on extroverted sensing and extroverted thinking (practical, results-orientated, efficient). Customer-focused or innovative organisations may be biased more towards extroverted feeling and both extroverted and introverted intuition. The other three introverted functions – feeling (employees' welfare), thinking (logical internal systems) and sensing (practical details of day-to-day internal management) – will all feature to some extent in the profile of most organisations, but it is doubtful whether the majority of companies would ever regard them as a priority over marketplace success.

The simple truth is that many individual preferences are essentially incompatible with the *overall* focus and aims of the organisation. The important thing is that the individual finds a positive expression for his or her preference within the context of his or her work. We have already seen how many externally focused companies rely on introverted types for the generation of ideas and the evaluation and monitoring of systems – the trick is to convince the organisation that shutting yourself away in your office for weeks on end is making a meaningful contribution to the bottom line!

Working within a team

A good team, like a good organisation, needs to take account of issues both external and internal to the team, and be both task- and people-orientated. Were Jung to construct the ideal team, we rather hope he might comprise it of eight individuals, each with a different and strong preference for one of the following:

- *Introverted Intuition*, to generate pure creative ideas that the team may (or may not) be able to work with.
- *Extroverted Intuition*, to see a relevance, meaning or pattern in the idea, and to identify an opportunity where it might be used.

- *Introverted Thinking*, to evaluate the idea objectively against independent criteria.
- *Extroverted Sensing*, to find a way of turning the idea into practical reality.
- *Extroverted Thinking*, to set a framework within which the team must achieve results, and to control progress against a plan.
- *Introverted Sensing*, to check that all the necessary details are being covered, and that deadlines are being adhered to.
- *Extroverted Feeling*, to keep the team united behind its purpose, and to liaise with people outside the team.
- *Introverted Feeling*, to make sure individuals in the team are happy, and to keep an eye on interpersonal difficulties.

We have no record of Jung's success in the field of teambuilding, but then, he didn't have the benefit of the questionnaire at the front of this book. Figure 7 shows the ideal team on the quadrant.

Those readers who are familiar with the work of Dr Meredith Belbin on teams may notice that, although Belbin does not refer to Jung in his work, it is possible to find a significant overlap between the functional preferences listed above and Belbin's identified roles of (in the same order) Innovator, Resource Investigator, Monitor–evaluator, Company Worker, Shaper, Completer–finisher, Co-ordinator, and Teamworker (his subsequently identified ninth role, that of Specialist, could conceivably be filled by any of the eight Jungian preferences).

It is important to be clear about your own preferences when working in a team, not only so that the team can get the best contribution from you, but also so that you can get the best from the team. Question 16 of the questionnaire lists eight qualities which all teams should have. Here they are again, with their respective preferences revealed:

- The team has a clear practical purpose (extroverted sensing).
- It makes the most of every opportunity (extroverted intuition).
- It is achieving results (extroverted thinking).

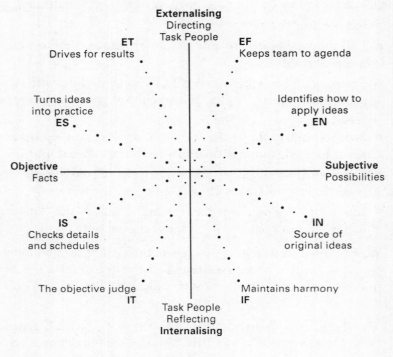

Figure 7

- It has good links with the people outside it (extroverted feeling).
- It completes tasks efficiently and well (introverted sensing).
- It finds new and creative ways of doing things (introverted intuition).
- It considers things logically and in depth (introverted thinking).
- Its members are happy (introverted feeling).

Depending on the functional preference of the person doing the measuring, a team's success can clearly be measured in many different ways, not only by those within it, but also by the organisation. A company with a strong extroverted thinking bias may lose patience with a team that is heavily introverted and intuitive unless the third quality listed above – tangible achievement – is present. For the sake of internal politics, if for no other reason, it may well be worth memorising the list.

Managerial styles

One can appreciate why the Eskimos have countless different words for snow when one considers the term 'manager', which is a hopelessly inadequate summary of the wealth of experiences and approaches embraced by this country's several million managers. The word tends to conjure up our image of the stereotyped line manager, constantly monitoring, checking and enforcing the contents of the several hundred memos which pass across his (for this stereotype is always a 'he') desk every day. This image may well be music to the ears of those with a preference for the extroverted attitudes of sensing and thinking – although this style

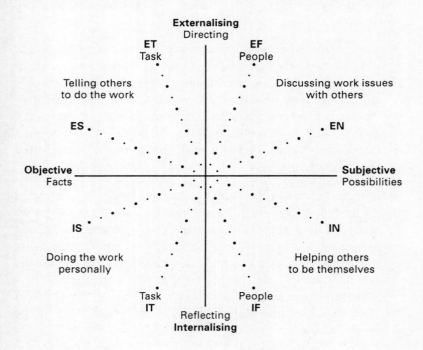

Figure 8 *Management styles*

of management, like most things in business, has been developed comprehensively from the bare bones which form the description above – but the fact is that anyone, with any of the eight preferences, may well find themselves in a managerial position. Needless to say, different people will approach the job in rather different ways. Figure 8 shows the individual preferences in managerial context.

The descriptions in this section deliberately make the false assumption that each of us uses only one preference out of the eight. This is clearly untrue, and goes against the grain of what this book is about. We have nevertheless taken this approach in order to give some clarity to the effect of each preference on managerial style. The authors have found in reality that most people in managerial positions find some valid way of using extroverted styles, since unless one is highly self-aware, the introverted preferences can be counter-productive in the social environment of management.

- The *Extroverted sensing* preference leads to a management style focused very much on the practical, present-day needs of a situation. This is the troubleshooter manager, quickly appraising a situation and responding appropriately. Because sensing is a perceptive style, a manager using this preference exclusively may be weak on closing off a project or following through, preferring instead to concentrate on the 'firefighting' aspects of the job.

- *Extroverted intuition* produces a manager just as responsive as the firefighter described above – but this manager is responsive to possibilities rather than actualities. A manager using this preference is likely to be innovative in approach, always looking for new ways to improve performance in the future, and able to respond flexibly to ideas for changing the way things are done. Closure and follow-through, again, may be a weakness, and practical details may be overlooked in the pursuit of ideas.

- The *Extroverted thinking* manager is not a million miles away from the stereotype portrayed earlier. This preference, with its love of systems and monitoring, is likely to lead to a managerial style which focuses on task control, performance against plan, and a generally organised and principled approach to the matter in hand. Such a manager is likely to be scrupulously fair in his or her dealings with subordinates – as long as they are straight in their dealings with him or her – but may have a problem dealing with people issues, because the temptation with extroverted thinking is to convert everything to impersonal data for evaluation.

- *Extroverted feeling*, on the other hand, usually values people first and foremost, and therefore will tend to produce a manager with a clear people focus. This might be demonstrated in extensive liaison to discover what external customers and the staff require from a department, and will certainly mean that those working for such a manager will be made aware of their value as individuals. They may be consulted on decisions which affect them personally, and the manager will spend a lot of time ensuring that the working environment is as harmonious as possible. The danger for such a manager is an over-reliance on values – both the manager's and the subordinates' – which may obscure logic at a time when it is needed for certain decisions.

- The *Introverted sensing* manager is as potentially adaptable as his or her extroverted counterpart, but given the reflective nature of introversion, this manager will be less of a firefighter: just as down to earth, but more pragmatic than practical, more realistic than active. This manager will focus very well on details, but may fall into the trap of 'nitpicking' without providing any constructive solutions. He or she may also end up doing a lot of the work personally, being far better at seeing what needs to be done than at communicating this to subordinates.

- Given the imaginative nature of *introverted intuition*, it is no surprise that a manager relying on this preference is likely to be highly individualistic and insightful – and almost certainly unconventional. His or her focus may well be on giving subordinates the opportunity to express their own individuality and creativity – but probably without any practical focus. Given that this is the most detached and subjective of all the eight preferences, any manager operating in this way is likely to encounter far more practical problems and opposition, unless he or she works for a very idiosyncratic company.

- *Introverted thinking*, as we have seen in other cases, has a very different flavour from its extroverted equivalent. Whereas extroverted thinking managers tend to be adept at driving for results, the introverted thinking manager will focus more on analysing the results themselves. He or she will have the same enthusiasm for efficiency, but is very unlikely to lead from the front in the way that extroverted thinkers do. However, the clarity of thought provided by introverted thinking will always be a valuable asset when it comes to seeing the best way out of a problem – even if the manager is more likely to enjoy the process of analysis rather than its outcome. The introverted thinker is also twice removed – once by the lack of feeling and once by introversion – from any natural adeptness at dealing with people, so people issues are likely to remain untackled.

- The *Introverted feeling* manager is likely to be highly receptive to any interpersonal or people issues among his or her staff, but will tend to unite the group not from the front, but by his or her presence in the background, wearing a mantle of passive accessibility. A manager using this preference may not provide much direction, but will always be full of encouragement if a subordinate comes to him or her with a difficulty or a new idea. Such a manager will probably praise before criticising, but may well encounter problems in pushing through difficult decisions or when discipline is needed. That having been said, people with this preference are often good at learning assertiveness techniques, because of their insight into people's feelings and how to use them positively.

Problem-solving styles

As with management, most popularly held views of problem

solving equate the process with what Jungian theory classes as extroverted sensing and extroverted thinking – logical, step-by-step approaches to solving a problem. This approach, of course, is not to every problem-solver's taste, but its prevalence as a prescribed 'textbook' method has probably led many individuals with preferences in other areas to conclude that they are no good at it.

The problem-solving styles, strengths and weaknesses suggested by each preference are described here and summarised in Figure 9.

- The **Extroverted sensing** approach to a problem is resolutely practical. The ES method is to look at the facts and find a way to take direct practical action – the method used will depend on the precise nature of the problem since sensing, being a

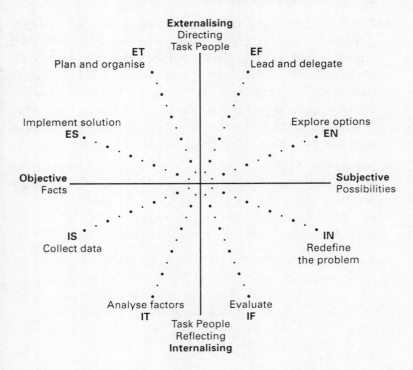

Figure 9 *Approaches to problem solving*

perceptive function, adapts to reality. Any solution found using this preference is likely to be the most practical, realistic and economical one available.

- **Extroverted Intuition** is equally responsive – not to the nature of the problem, but to the nature of the options available to solve it. Someone with this preference is likely to concentrate on generating as many solutions as possible, relying on someone else to fit the correct one to the problem.

- **Extroverted Thinking** looks for the most coherent and logical plan of action for dealing with the problem. This function is ideal for fitting the facts – or possibilities, for thinking is as often teamed with intuition as with sensing – into an orderly and objective system, so that the problem can be dealt with in as efficient and trackable a way as possible.

- **Extroverted Feeling** is the preference most valuable for group problem solving. It performs the same function as in teamwork generally: uniting the group behind the task in hand. Someone with this preference will not only perform a useful function in motivating the members of the group – or, if a group does not exist, in enlisting the help of others to solve the problem – but will also be able to weigh up the value of the potential solutions in terms of their likely benefit to people.

- **Introverted Sensing** focuses on the gathering together of relevant facts, on finding out as much as possible about the problem. Someone with this preference will tend to reflect on the data they are collecting, perhaps adding a personal interpretation to push the process onwards.

- **Introverted Intuition** is the source of the problem-solving technique known as 'reframing' – finding a new angle from which to look at the problem, or seeing different aspects of the problem which may enable an unexpected solution to be found. Often the greatest insights into problems come from someone with this preference.

- **Introverted Thinking** will evaluate the problem using the thinker's own system of logic. The problem will be examined and analysed in terms of its relationship to a conceptual model, and potential solutions will also be thought through in a deep, systematic way. This preference can also supply objectively critical, sometimes sceptical reactions to proposed solutions.

- **Introverted Feeling** will also evaluate the problem and potential solutions, but in terms of how others feel about them. Someone with this preference, however, will very likely not consult those people affected, but will rely on his or her instinctive empathy to put himself or herself in the others' shoes. This preference is also more likely to take into account people's concerns about the issues involved.

Unless you happen to be solving problems in a perfectly balanced team of eight, there is a definite advantage in learning to use – or at least to appreciate – all the functions when called upon to solve a problem. At the very least, recognising which preference is most likely to bring results in any particular case can be invaluable when enlisting the help of others in a work setting.

Communication styles

Communication, like many other activities in our society, is at its best when it takes place with the informed consent of both parties. It is at its worst when it is taking place more for the benefit of one party than the other. Good communication, therefore, is a complex skill because it requires not only an awareness of your own strengths and weaknesses, but also an insight into those of the person with whom you are communicating. All too often, people in organisations treat communication as a one-way activity – which goes some way towards accounting for the almost universal unpopularity of memos.

The two aspects of communicating – telling (or giving out the message) and absorbing (or receiving the message) – neatly parallel the extroverted and introverted sides of Jung's model. Given, as we have observed already, that the average individual tends to behave in a more extroverted way at work, it is not surprising that listening skills are often those least in evidence among the average work group.

There is, of course, fault on both sides. Somebody with an introverted bias to their profile is quite likely to be an excellent listener – but they may hold back from opportunities to give out messages. These are the people who will send somebody a fax rather than ring them up, or circulate notes rather than canvass opinion personally. Extroverts, on the other hand, may be so keen to influence the environment around them that they find it more difficult to stop and take in properly the messages they are receiving from the other party.

Each of the eight preferences carries with it a distinct communication style (Figure 10) – and an equal capacity for crossing wires should the recipient of the communication be using an incompatible method.

- **Extroverted Sensing** involves concentrating on giving a clear outline of the facts. Someone with this preference is likely to be

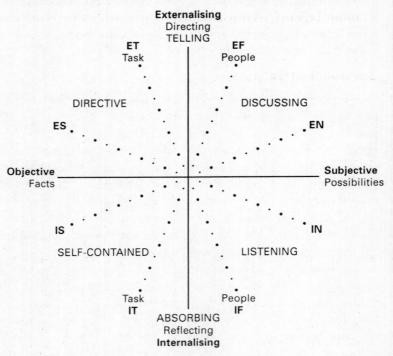

Figure 10 *Styles of communication*

good at sticking to here-and-now, 'live' issues in their communication with people. Listening to them, you are likely to know exactly where you stand. If they are trying to negotiate with you, they will propose tangible concrete benefits, and paint a clear verbal picture of the situation as it stands. As listeners, they tend to be good at giving a clear accurate summary of what they have heard, but may not wait as long as they might before doing so.

- **Extroverted Intuition** leads to the communication of a clear vision – someone using extroverted intuition will invite you to share in a picture of possibilities, of new views and ideas. They may well find it hard to stick to the same subject without going off at tangents. The language of their written communications is likely to be highly embroidered and original. When negotiating, they tend to talk in broad terms about possible future benefits. As listeners, they look for patterns and meaning in what they are being told, and may be adept at pulling together another person's thoughts.

- **Extroverted Thinking** leads in communication to a focus on presenting clear logical reasoning. People with this preference organise their communication – verbal as well as written – in an orderly, objective way. Their arguments are likely to be coherent and well thought through. In negotiating, they will concentrate on giving logical, principled arguments for the benefits of what they stand for. As listeners, they have an ability to present back to the speaker an ordered, organised version of what he or she has said – but they are likely to be impatient and to interrupt a rambling argument.

- **Extroverted Feeling** provides the ability to express one's own feelings clearly to others. People with this preference may well be the best of all at talking to someone 'in their own language'. Their ability to make contact with people means that they will probably choose verbal communication whenever possible, and will tend to base their reasoning on what they get back from the listener. As negotiators, they will concentrate on selling the people the benefits of their solution. As listeners, they are likely to be good at making the speaker feel attended to, but their warmth may well encourage them to join in, finishing off the speaker's sentences or anticipating his or her words.

- **Introverted Sensing** concentrates on distilling the essential details from a communication. People with this preference can excel at the art of précis, and their communications may well be briefer than those of someone with an extroverted sensing preference. Written communications from them may even seem curt, as only the essentials are deemed to be necessary for inclusion. As negotiators, they rely on influencing others by giving their own impression of the true facts of the matter. When listening, they concentrate on absorbing the detail of what they are hearing, and can pick out the one or two key points from what was said.

- **Introverted Intuition** is likely to lead, in those with a preference for it, to unconventional and imaginative ways of communicating. The method may not be new, but their use of the method may be – memos from someone with this preference may well break all the unwritten rules of company memos. They often do not worry too much about how the recipient(s) of their communications react to them, as their reward lies mainly in the process, not the ends. In negotiating, they often rely on either reframing the situation to find a new, unseen benefit or, if the benefit is clear, on focusing on its possibilities in specific situations. As listeners, they will again tend to offer back a new angle on what they have heard – after they have chewed on it for a while.

- **Introverted Thinking** weaves complex theoretical models into its communications. People with this preference are far more likely to communicate in writing than verbally, and what they write will take some time to absorb, since it can be overwritten, with long sentences constructed around central logical structures. In face-to-face communication they are at a disadvantage, unless their profile contains some degree of extroverted feeling, since without it they are tempted to remain impersonal and to continue to present complex arguments. As negotiators, if they curb their love of complexity, they are good at giving an in-depth analysis of the key factors being discussed. As listeners, they can be excellent at absorbing the tangled logic of what is said to them, and untangling it for the speaker.

- **Introverted Feeling** imparts to its owner the temptation not to 'communicate' at all, but rather to simply co-exist, as it were: for people with a strong introverted feeling preference are so clear about their understanding and empathy for people that they do not always feel a great need to communicate directly with others. When they do communicate, they are very careful how they choose their words, in order not to disrupt the potential harmony between themselves and the recipient. Whereas extroverted feeling would be more likely to re-establish harmony after a misunderstanding, introverted feeling may well withdraw from any conflict. As negotiators, they can be supremely responsive to the moods and signals of the other party – as is the case when they listen. This is the typical counsellor's preference, and its capacity for absorption and empathy usually makes excellent listeners.

Time management

The effective management of time is another of those areas for which only one approach seems to be generally accepted as being the 'right way'. Most of those people who regard themselves as poor time managers blame it on an inability to schedule correctly, to control and monitor their own timekeeping. By this point in this book, it may be stating the obvious to point out that the success or failure of one's personal approach to the management of time depends ultimately on the efficiency with which one can apply one's own personal preferences to the task. Figure 11 shows how individual preferences affect attitudes to time management.

- The **Extroverted Sensing** approach to time management is essentially realistic and reactive. Time is simply a measure of how long it takes us to do things, and provided we approach each thing we do with practical efficiency and speed, that

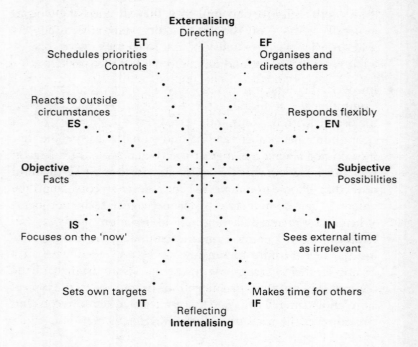

Figure 11 *Approaches to time management*

length will be as short as possible. Extroverted sensing does not encourage the use of planning techniques for time management, since sensing as a function ultimately relies on the assumption that life cannot be scheduled or predicted – it can only be responded to.

- For **extroverted intuition**, time is a resource to be used fully. The goal is to explore all the possibilities in the time available. Scheduling, again, is counterproductive, and a dangerous temptation for someone with this preference who, in any case, would always rather be living in the future than the present. Extroverted intuition works best when it can concentrate on the matter in hand, but also be free to ask the question: 'How much is it possible for me to get done in the time available?'

- As in many other areas of management, **extroverted thinking** is the archetypal time-management approach. It is orientated to performance against logically set deadlines: each task is assessed and fitted into a schedule, with all necessary logical contingencies attached. The unpredictability of life, which sensing and intuition regard as unplannable for, is merely another set of data for extroverted thinking: unpredictability can be built into a schedule just as easily as anything else.

- **Extroverted feeling** recognises the difficulties of scheduling the unschedulable, but carries with it the confidence that the most unpredictable of time-management factors – people and their values – can be managed. This preference may not be as orientated towards schedules as extroverted thinking, but nevertheless will prefer to operate from a schedule of some kind. The schedule, however, can be suspended when, for example, the interruptions of others make it difficult to follow. Whereas thinking allocates time to tasks on the basis of how much time it is logical to give them, extroverted feeling allocates on the basis of what feels right, or even how much time each task deserves.

- As with extroverted sensing, **introverted sensing** carries with it no attempt to plan or think out use of time. Rather, people with this preference absorb facts and details and then spend their time on specifics, often losing sight of the broad picture – what achievement is being aimed for. Their strength lies in completion, finishing details and focusing on the particular. Time can often lose all meaning for them as they lose themselves in dotting the i's.

- Time has little meaning also for those with a preference for **introverted intuition**: it is regarded as a convention to be risen above. This does not meet with most organisations' views of good time management, but someone with this as a main preference would not really worry about that – his or her chief strength would be an ability to work intensely when motivated, and if the motivation wasn't there, the task would not be worth giving time to in the first place.

- **Introverted thinking** regards time as the ultimate limitation on what can be done – it cannot be overcome, so it must be fitted into a conceptual model. Introverted thinking finds it very difficult to be flexible once time has been allocated. Its extroverted counterpart carries a greater ability to organise events to fit the schedule, but introversion tends to work against that, relying on the outer world to fall into the pattern originally scheduled. When it does not, introverted thinking is helpless – unless a new logical schedule can be drawn to fit the new version of events.

- **Introverted feeling** schedules on the basis of people's needs – either the manager's own or other people's. Again, introversion tends to work against the capability to influence events successfully, so there is less people management involved than would be the case with extroverted feeling. Instead, more contingency tends to be built into the schedules of someone with this preference – based very much on their own weighing up of the value of the task.

4. What does it all mean?

Work compatibility

For many people, the work that they do and the organisation within which they do it are bound together so that there is a blurred zone where the two meet. We are treating them separately here, because it is fundamentally important that anyone considering their place within life's rich tapestry – or at least whether they're in the right job – is able to draw a distinction between what they do and where they do it. We have worked with individuals who have been desperately unhappy and on the point of leaving a firm to go somewhere else, who have suddenly realised that what they actually need to do is stay where they are but change their job.

The great sadness is that this course of action is a luxury not available to all: very many organisations are rather inflexible when it comes to lateral moves; but many more, especially in the current decade, are waking up to the fact that it is better to keep a loyal employee but lose the work they *were* doing, than to lose the employee *and* the work. But even in these companies, you are unlikely to find such adaptability spelt out in the company handbook. It *will* cause them upheaval, and consequently they may not want to overencourage lateral moves. However, if you are unhappy with the work you do, but happy with the company you work for, it is always worth approaching them about the possibility of a change of job within the organisation before you draft your letter of resignation – you have, after all, very little to lose.

Modern management development is revolving more and more around two sacred cows: competency assessment and training needs analysis. The former aims to identify with needlepoint accuracy the particular qualities and abilities required to do a particular job in a particular organisation; it implies that an individual picked at random will find somewhere in the world a job with 'their name on it' – a perfect match for their unique make-up. The latter aims to ensure that an individual already doing a particular job acquires all the skills they need to do the job as best they can. Both work on a kind of identikit basis: the former looking for a perfect match before the event, the latter making up for any shortfall after the event and before it's too late.

The concept of preference which we are exploring in this book, and the functional model on which it is based, tend to defy such a rigidly scientific approach. Undeniably, it overlaps with the two methods described above: show us a list of competencies identified for a particular job, and we could fairly reliably predict the functional preference of the individual who eventually turns out to be the ideal incumbent; look at a profile of a department's training needs, and it is possible to identify particular functions the team may be deficient in. But the point which cannot be ignored when dealing with individuals in individual jobs is: *it is perfectly possible to be thoroughly unsuitable on paper for a job and still enjoy it and still do it well*. Provided you are being true to your own preferences, you can adapt (within reason) the way you approach a job to suit both the job and you.

If we take a broad managerial role, such as sales manager, for example, we can look at the effects of different preferences on the job performance. The obvious preference for a sales manager to have, one might assume, is extroverted feeling, since the job must involve dealing with people, understanding their values, and focusing on their needs. Well, maybe: but the functional opposite, introverted thinking, could also be very productive in a sales manager. It would produce a very different kind of employee: this sales manager would be likely to shut himself or herself away in an office looking at lists of sales data, and analysing whether the efficiency of the sales team could be increased, for example. A sales manager with a strong extroverted intuition preference

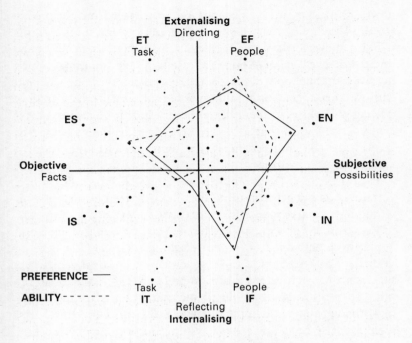

Figure 12 *Pamela's profile*

would be far more interested in dreaming up new imaginative
sales campaigns and getting them out on to the street. And so on.

There is not room in this chapter to go through the effect of
every possible functional preference on every possible occupation.
We have a few genuine (but disguised) case studies from our work
with individuals which will make the concept clearer. In
describing them, we also acknowledge that it is certainly not
always possible to adapt completely to a role which does not fully
suit one's preferences.

Pamela is in her mid-20s. She currently works as assistant
marketing director for a professional association whose members
are all accountants. As such, she is effectively responsible for
letting everyone know how wonderful accountants are. Never

having worked as an accountant (and, as you see from the profile (Figure 12), having fairly low scores for sensing and thinking) this is not a message that comes straight from her heart. What she loves, however, is marketing. Or selling. Or anything which involves finding new ways to make contact with people and enthuse them. She originally worked in recruitment, mainly dealing with the placement of temporary staff for a major central London agency. In that job, her extroverted feeling was used to the full, as she charmed her way down the telephone lines of various large companies.

Recruitment, however, paled after a while. Pamela's profile, you notice, has a generous amount of extroverted intuition. This would have been demanding of her constantly that she set up new projects, dream up new and different ways of doing things, and apply those new concepts to the world around her. It has to be said that working in a 'temp' agency is not an ideal vehicle for such a strong extroverted intuition preference. For a start, in the fast and furious environment of making and confirming bookings, ringing temps, logging all the details, checking timesheets, what is really needed is a strict system, not a policy of constant change and new ways of doing things. In fact, the company had a rigid system. It would have ideally suited someone with as strong an extroverted sensing score as Pamela had for extroverted intuition. Sensing would have taken care of the office reality, while feeling took care of the temps and the clients.

So Pamela had to make a move into a profession that retained her need to reach people and influence them, but replaced an emphasis on factual reality with an emphasis on new ideas. The result: marketing.

But Pamela, unfortunately, has not found her El Dorado. The process of marketing satisfies her extroverted feeling and extroverted intuition; but there is one other significantly high score on her profile with which we have not dealt: her introverted feeling. This, you will remember, is all to do with personal values, with empathy, and with depth of personal feeling. This score suggests that for Pamela, it is very important that her work gives her a deep feeling of reward, and unfortunately marketing the virtues of accountancy does not push any hot buttons (as a marketing person might put it). Put simply, the way that she approaches marketing is consistent with her profile, and makes her happy; but *what* she is marketing is not consistent with her profile, and so leaves her feeling 'Is it worth it?'

Had Pamela a stronger score on introverted intuition, perhaps she might be able to reframe the situation and change the emphasis so that the accountancy aspects did not loom so large. Had she stronger scores on introverted sensing or introverted thinking, she might understand more what accountancy is all about, and her natural capacity for empathy (introverted feeling) might lead her to identify more with those she serves. She not only lacks these, but her extroverted feeling is also partly frustrated. This function not only leads her to build harmonious relationships with clients and external agencies (which is fine), it also gives her a drive for affiliation within the office. Since most of those she works with are accountants (including the marketing director – her direct superior) she is stymied. At the time of writing she is looking for another job.

Anne is a teacher. She has worked in both the state and private sectors, teaching English, French and drama at secondary level. Her main love, however, is for drama, which she regards as a subject which ought to be at the heart of every school's curriculum. She is fortunate enough to work in a profession which has scope for individual approaches, and her approach can be guessed at from her profile. You will see that by far the major share of her total score is split between introverted and extroverted feeling (Figure 13). Introverted feeling gives her a passionate belief in her subject and in the importance of what she does, and gives her a strong empathy with those she teaches. Extroverted feeling allows her to reach those people, and not only to reach them, but to form strong and highly personal relationships with each and every one of them. This is the bedrock of her teaching style: she has a limitless capacity to enthuse her students and to make them feel special.

Once feeling is accounted for, the other scores are left rather behind, but there are two others with some significance in this profile: extroverted sensing and extroverted intuition. These two

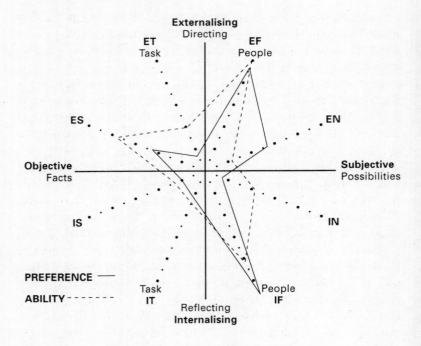

Figure 13 *Anne's profile*

functions provide the raw material which Anne's feeling converts into a learning experience for her students. Intuition provides the imaginative input on which good drama relies: the ability to devise role-play scenes which embody a particular concept of life skill; the ability to place students in roles or situations which particularly suit them. Sensing provides the background of reality which Anne regards as crucial if drama is to have a meaningful place in the school curriculum. She takes great pains to ensure that all her lessons have an application in the real world, and that the imaginative leaps of which drama is capable are always firmly anchored to the ground.

It seems like the perfect profile. But teachers like Anne have their cross to bear: Anne works in a profession which is regulated by many and varied sets of rules and requirements. These are obviously there for a purpose; but teachers like Anne, whose success lies with her unique ability to form productive relationships with individuals, tend to feel restricted by, for example, grading structures which require certain probationary periods in certain types of school. (She worked part time originally, taking time off to have a baby, and discovered that, after four years during which she had effectively acted as head of a department, she was still regarded as not having finished her probationary year.) Diagnosed as innumerate while herself at school, she discovered on first attempting to get work that without a maths O-level, she could not obtain a DES number (essential to work in the state system) – even though she was never going to teach maths as a subject.

Anne is someone whose profile is dominated by feeling – which defies any attempt to set up a logical system – and is low on thinking, so it is not difficult to imagine how impassioned she can become when discussing what she sees as a conspiracy to stop her doing what she is good at. This is also a good example of a situation in which changing organisations will not work – although some schools are more tolerant than others, professions such as teaching carry a pull from the invisible administrative centre which seeks always to standardise (or to restrict, as Anne would have it).

What Anne does in response is, first of all, not to change jobs. She loves teaching with a passion (hardly surprising given her prominent feeling preference), and would not contemplate doing anything else. She also would not contemplate developing, say, her thinking function (she hates logic and objectivity with an equal passion). What she does instead is use her feeling function constructively to 'beat the system'. She ensures in every school she works in that she is never seen to rock the boat, or to openly defy any of the systems she hates. True to her preferences, before very long she becomes not only a teacher who always gets results, but one who is highly popular with everyone she teaches – she is frequently seen to have the most anarchic and 'unteachable' pupils eating out of her hand. She also charms most of those who pass her way, as do most people with a strong extroverted feeling preference.

Once her position is this secure, she can start bending the rules she dislikes, and rely on her popularity to see her through – the level of support she is bound to receive from pupils tends to ensure that any dissent is swiftly dissipated. She also uses her anger at what she regards as pettiness to fuel her driving feeling, and so ploughs it back into the job. Whereas Pamela's extroverted feeling cannot ultimately help her, because it cannot make her love the subject of her work, Anne already loves her subject and needs only to sell it effectively to those she works with.

Organisational compatibility

Having separated out the factors which influence our compatibility with the work we do, we need now to examine the other side of the equation – how we feel (or think, or sense, or intuit) about the organisation for which and within which we work. This means not only those intangible aspects of 'climate' and 'culture' much beloved by consultants, but simple facts about what the people you work with are like, what your own development prospects are, or whether your boss has halitosis.

When dealing with organisational characteristics in Chapter 3 we pointed out that it is possible to profile an organisation in the same way as an individual, using the eight Jungian functions. An individual with a strong extroverted feeling preference working for an organisation whose environment is more orientated towards extroverted thinking (organisation, systems, control, focus on results) may find they run into problems or conflicts.

Pamela and Anne, leaving aside the issues regarding their work compatibility, have both encountered this kind of conflict. Pamela's organisation, representing accountants, has a strong orientation towards introverted thinking (focus on internal systems and processes) which conflicts with Pamela's own strong extroverted feeling preference. In Anne's case, the school for which she works, as is the case with many schools, has a focus on achievement of results and control of its pupils' environment, which translates as an extroverted thinking profile. This, for Anne, represents the systems against which she spends much of her life fighting. Fortunately, schools tend to have a strong client (pupil) orientation and concern for personal development which results in a strong extroverted feeling preference, so she finds her match.

The key point about the match (or mismatch) between individual and organisation is that the organisation's orientation will effectively set the ground rules for the way work is done. It will determine the procedures to be followed and also the actions which will be rewarded. Large public organisations, for example, often end up with an overall organisational preference for internal systems (introverted thinking) and for conformity to detailed procedures (introverted sensing). In other words, such an organisation becomes bureaucratic. What is lost is sight of the external client (extroverted feeling), or getting resources and getting things done outside the organisation (extroverted sensing and thinking). One of the consequences of the government's privatisation programme was that companies were forced to develop their organisational extroverted functions.

The recent history of the National Health Service is a pertinent example. The 'internal market' of the NHS means that individual trusts have to become 'client focused' (extroverted feeling) and results-orientated (extroverted thinking). Many of them are looking at new ways of doing things (introverted and extroverted intuition) whereas before they were led by the precedents of laid-down procedure (introverted sensing and thinking).

Individuals within organisations such as these, depending on their individual preferences, will either welcome such changes or feel threatened by them. In a bureaucracy one is rewarded for attention to detail and procedures: in a customer-orientated or 'innovative' organisation, procedures are to be ignored if they get in the way of client service. So the individual will have to act in very different ways, regardless of his or her job, function or preferences, according to the organisation's preferences and values.

This explains why someone can move from doing a job not very well in one organisation to doing exactly the same job in another organisation and being brilliant at it. The person in question has not improved his or her abilities *per se*, but is able to use his or her preferred functions in a more compatible

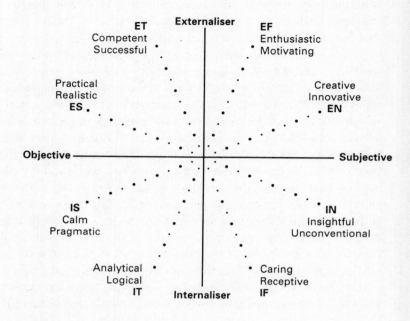

Figure 14

environment. Result: stress levels down, good performance, happy individual, happy organisation. So people need to make career decisions not just on the basis of the job they will be doing, but also very much on the type of organisation they will be doing it in.

Danger: people at work!

Infuriatingly for some of us, our offices seem to be full of other people. Not only do we have to share railway carriages with them, or sit next to them in traffic, we actually have to work with them as well. Many readers may be thinking that their jobs would be OK if it were not for the people they work with. This is the right moment to pause and consider exactly what type of people you might actually enjoy working with.

Before we continue, a word of caution. Even when someone is clear about his or her own profile, one man's meat, as we all know, may well keep his dinner party guest up all night. To put it simply, the theory that opposites attract runs up against the theory that birds of a feather flock together. One of the authors of this book married his functional opposite; the other married a Jungian clone. As yet, no petitions have been filed; babies have bounced and thrived in both houses. The summaries in the quadrant in Figure 14 describe ideal workmates, based on the premiss that the reader would like to work with people with the same functional preferences as him or herself. If any reader values difference more highly than mutuality, he or she will easily be able to extrapolate the information needed to purge the office of undesirables. Look back, too, through the descriptions of styles of working on pages 50–53. If there are any you particularly dislike, watch out for people who adopt them in your own environment. At least now you know they have an excuse – Jung says it's innate.

5. Copability

Satisfaction and stress

Stimulation is extremely important for us all. Lack of it produces boredom and fatigue, and the ensuing reactions are what we tend to recognise as stress. Over-stimulation can also produce stress reactions, brought on by feelings of not being able to cope. Our working definition of stress is: *The reactions of individuals to situations which are either overly or insufficiently demanding in terms of the individual's capabilities.*

Stress, like many unpleasant features of life, has a positive aspect: it can provide a signal to people either to improve their capabilities or to change their situation. Either way, stress demands some form of action.

Stress reactions

The stress reaction is seen by Hans Selye (the guru of stress) as a three-stage process within the body that enables it to survive and adapt to change. It is a throwback to our cave-dwelling days when, on being confronted with a sabre-toothed tiger, stone-age man would experience what is now termed the 'fight or flight' syndrome: a rush of adrenalin which would temporarily provide the alertness and energy either to attack the animal or to run away. It remains with us today, like the tiny tail at the base of our spines, a reminder of what we once were.

The first stage of any stress reaction is alarm. The surge of energy, concentration and power that comes with the stress alarm enables people to perform in a crisis – sometimes considerably beyond their normal physical capabilities. Another reaction at this stage, which Selye does not mention, is the 'freeze' response: we are rooted to the spot – in the hope that the sabre-toothed tiger will ignore us. This is called by some the 'possum effect'.

Once the alarm stage of stress has passed, the body enters a second stage, one of recuperation in which it repairs any damage caused by the demands of the fight-or-flight response. The third stage is a return to the body's normal state of relaxed alertness. 'Bad' stress is simply normal or acute stress that becomes chronic, continuing for weeks and months so that the body never gets time to recuperate.

Chronic stress can inflict real bodily harm – especially by lowering resistance to disease through the immune system. It is often cited as actually causing disease, but there is no real evidence to support this, beyond a possible contribution to ulcers and heart disease.

It is easy to identify isolated events which cause a big stress reaction; it is harder to look at the mental, emotional and physical habits that cause stress each day. These stretch us like a spring when a weight is put on it.

Hooke's Law of Elasticity provides a useful reference point for this analogy. It refers to the twin factors of 'stress' – the load (or demand) which is placed on a piece of metal, and 'strain' – the deformation which results. Hooke's Law states that if the strain produced by a given stress falls within the 'elastic limit' of the material, then the material will simply return to its original condition when the stress is removed. If, however, the strain passes beyond the elastic limit, then some permanent damage will result. This allows not only for the idea of a sudden immense pressure but also pressure which is more persistent and drawn-out. It also includes the notion of an instinctive resistance to some degree of stress. Chronic stress can produce deformation in us or even break us, just as a heavy weight can break a spring. We are all different, though, and have different reactions and different breaking points.

Sources of stress
Stress can come from either external or internal sources. External

(extroverted) sources can be environmental (such as physical working conditions) or interpersonal (such as the people you work with). Internal (introverted) sources include physical, bodily conditions, emotional reactions and needs, and states of mind such as inferiority complexes. As is the case with introverted functions generally, these internal conditions can be seen not just as reactions to the outside world but as entities in their own right.

The different sources can be mapped on to our quadrant model as shown in Figure 15.

Go back to the stress section of your questionnaire results (see page 26). Look both at the general profile and at your scores for questions 22 and 23, which cover what causes you stress and the effect stress has on you. As you read through the descriptions given below of the different functional reactions to stress, you may find descriptions which fit you but which appear to disagree with your

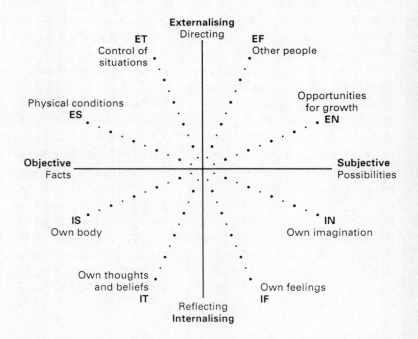

Figure 15 *Sources of stress*

profile. Note these down, bearing in mind that, as Jung saw it, stressful reactions often represent the preferences of our unconscious or 'shadow' self, and may differ greatly from the ways in which we usually prefer to work. We will return to this later in the chapter when we look at ways in which stress can be dealt with.

Internal stress factors
Research into highly anxious individuals shows that their thinking habits are often based on a negative view of themselves in comparison to others. They have a poor 'self-image'. In some individuals, anxiety manifests itself in emotional habits such as fear and guilt; others experience stiff necks or tense shoulders. If we experience stress internally, it will take one of these forms – mental, emotional or physical.

Mental habits (introverted thinking, introverted intuition)
Stress which involves the functions of introverted thinking or introverted intuition relies on our special private interpretations of events which determine our emotional response. The meaning of an event is embodied in a thought or image; if reality is distorted as a result then our emotions respond to the distortion rather than to factual reality. People who, in favourable conditions, have a preference for extroverted sensing or extroverted feeling will often react to stress through the (for them) unconscious and underdeveloped functions of introverted intuition or introverted thinking. Negative thoughts and images appear spontaneously; they usually involve considerable distortion of reality and are difficult, sometimes impossible, to shut out.

These are typical statements that might be made by someone experiencing stress through introverted intuition:

— *I imagine terrifying scenes.*
— *I can't keep anxious pictures out of my mind.*
— *My life lacks meaning.*

Typical statements resulting from stress in the area of introverted thinking might include:

— *I can't make up my mind fast enough, so I feel I'm losing out.*
— *I can't keep anxious thoughts out of my mind.*
— *I find it difficult to concentrate because of uncontrollable thoughts.*
— *I worry too much over things that don't matter.*
— *Some unimportant thought keeps bothering me.*

Emotional habits (introverted feeling)
These involve emotional tension and feelings such as panic and worry. Whereas introverted intuition raises fears of an event's possible unpleasant consequences, and introverted thinking provides a 'logical' argument as to why everything is going wrong, introverted feeling concentrates on the way the person feels during the stress. It may give rise to a feeling of worthlessness, of paranoia that everyone is commenting on the person's performance, or an emotional outburst.

Typical statements made by someone reacting in this way include:

— *I feel a greater dislike for people in general.*
— *I feel depressed and upset.*
— *I want to shout out.*
— *I want to cry.*
— *I get angry and irritable.*
— *I lose my temper over trivial things.*
— *I feel emotionally out of control.*

Physical habits (introverted sensing)
Much stress is physical. We are continually in fight, flight or freeze situations when the adrenalin builds up and we do not know how to discharge it. Most car drivers experience it at some time or another. Constant arousal leads to muscular tension which may gradually become a habit. Muscles lose their flexibility and act like armour to protect the body, thereby failing to work independently. Shallow breathing leads to a pent-up feeling of anxiety.

Typical statements resulting from stress linked to introverted sensing include:

— *My heart beats faster.*
— *I perspire.*
— *I feel shaky.*
— *I get diarrhoea.*
— *I pace nervously up and down.*
— *I become immobilised.*
— *I feel tense in my stomach.*
— *I can't relax physically.*

With these internal stress factors, as with all the introverted functions, it is sometimes difficult to distinguish between cause and effect. For some people, a distressing thought (introverted thinking) may cause a tense feeling in the stomach (introverted sensing); while for others, tension in the stomach caused by some physical illness may express itself as stress by provoking uncontrolled and unpleasant thoughts about the nature of the illness.

External stress factors
It is important for us as individuals to feel that we have some impact on the environment and on other people. When this outside world is overwhelming us and we feel we have no choice or ability to do anything about the demands being made on us, the result is stress.

Work and stress (extroverted thinking, extroverted sensing, extroverted intuition)
Work is one of our major activities in life. If it is satisfying it will have a positive effect on other areas of our lives; if it is not right then it can become a major source of stress.

It is obviously crucial that the personality and skills you have are right for the job. A simple example of mismatching would be someone who was outgoing and liked people having to work in an office all day by themselves. Stress would be the result. If you are well matched with the job you do, it is unlikely to be stressful.

Figure 16 shows an approximate relation between an individual's needs for challenge at work, and the demands of the job.

The lower shaded area represents those not particularly

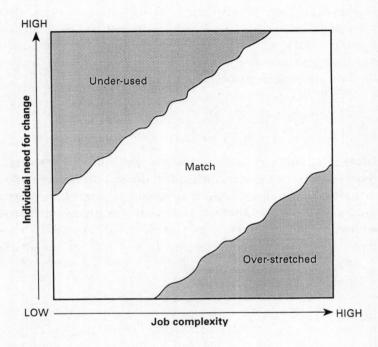

Figure 16

ambitious but doing difficult jobs. Stress occurs here because of 'over-stretching'. The middle area represents a happy match of individual need and job complexity. The upper shaded area represents individuals who feel under-utilised. Stress will also occur here but it is more difficult to identify, since the sufferer is likely to feel 'on top of' the job rather than submerged beneath its demands – but the boredom and lack of self-worth which come from being under-stretched can be just as damaging. In both stress areas there is a clear need to move into the middle. If you are over-stretched, training or changing the job may help. In the case of under-use, promotion or job enrichment may be the answer. If not tackled, demotivation can occur.

The stress section of your score sheet will give you some clues to the directions from which stress comes in your job. Some are easier to identify than others.

Stress in the area of **extroverted sensing** will be associated with the practical realities of your workplace: not so much physical comfort (which is closer to introverted sensing) but the concrete demands your job makes on you. Someone experiencing stress in this way may make the following statements:

— *I find it hard to achieve practical objectives.*
— *I can't cope with this mountain of paperwork.*
— *I lose track of where I am in the job.*

Extroverted thinking causes stress in similar, concrete ways, but, because it is more concerned with decision-making whereas extroverted sensing is concerned with perception, someone experiencing stress in this area may make the following distinct kinds of statement:

— *I can't organise my work properly.*
— *I find it hard to make an objective decision.*
— *I keep missing deadlines.*

Extroverted intuition is concerned with the possibilities inherent in situations, and its related stress follows a similar pattern, resulting in the following kinds of statement:

— *I can't cope with a new situation.*
— *I find it difficult to come up with new ideas.*
— *I am concerned about where my job is going.*

Stress and others (extroverted feeling)
Dealing with the physical realities of the job is one source of stress in the workplace: the people around you are another potent source. The following statements might be made by someone experiencing stress in the area of **extroverted feeling**:

— *I never know the right thing to say in a situation.*
— *I can't cope with people interrupting me all the time.*
— *I always seem to be being persuaded to do things against my will.*

With the extroverted functions it is easier to see how our inability to deal with certain kinds of situation may be a result of our not being developed enough in one or more of the functional areas. Someone with a high score on extroverted intuition, for example, should have very little trouble generating ideas – but it can happen! If you scored highly in the stress results in an area which is identified overall as a strongly preferred function for you, take comfort from the knowledge that you should find more tools at your disposal for coping with the situation.

Coping skills

Coping is an attempt to master a demanding situation which is seen as stressful. The more coping skills that are available, and the more confident people are in using them, the less likely is stress to lead to strain. Coping skills are affected by:

- a sense of personal control;
- adequate information;
- social support.

These factors increase a person's ability to feel that they can cope and are confident of meeting challenges. A key point is that 'good' coping means being able to shift from one appropriate way of dealing with stress to another, as the situation demands. People need to develop a range of successful coping skills. If they have a limited range or are not able to apply them successfully, then stress will turn to strain. In the long term this can lead to hopelessness, helplessness and depression. Obviously, your functional preferences will have an impact on the ease with which you can adopt certain styles of coping.

Ways of dealing with stress

Coping styles can be classified as expressive (direct action) or inhibitive (palliative) styles. Successful coping does not always involve active mastery over the environment: inhibitive attempts such as retreat, tolerance or disengagement may be the most healthy response in certain circumstances. In non-Western cultures, for example, habitual fatalism can serve a positive function in coping with stress.

Expressive styles involve dealing with the source of the

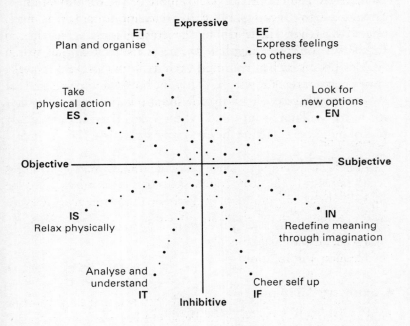

Figure 17 *Coping styles*

problem. The critical aspect is identifying the right target. If it is wrong, the stress will continue and, to make things worse, you may be left feeling guilty at having attacked the wrong person or thing. Assertiveness techniques can be of great use here. If the target is too strong, frustration may build up which can only be avoided by displacement – which roughly translates as 'kicking the cat'.

Inhibitive strategies aim to deal not with the stress source itself but with its effect on us: we have a choice of withdrawing from the stressful situation or denying its existence. Other strategies which fall into this category are the use of alcohol, tranquillisers and sedatives, meditation and muscle relaxation. Problems arise with a strategy of this kind when the person employing it fails to 'return to base': an example of this would be someone choosing to withdraw from a situation in order to cope with emotional reactions caused by stress, but then preferring to remain in that state rather than returning to the fray, batteries recharged. Withdrawal or denial of stress can be of use in saving us from everyday hassles such as commuting to work. It can break down, however, in extreme cases.

These two types of coping style mask a variety of responses, which, you will not be surprised to hear, map on to the functional areas of our model. Taking the expressive styles first, they include:

- **Extroverted feeling**
  Expressing your feelings
  Letting off steam
  Talking to others
  Confronting others with the truth of a situation
  Getting advice
  Changing relationships

- **Extroverted intuition**
  Looking for new ways of doing things
  Looking for new opportunities
  Experimenting with different ways of dealing with a situation
  Exploring options

- **Extroverted thinking**
  Analysing and getting rid of the sources of stress
  Setting priorities
  Planning and organising
  Getting in control of a situation

- **Extroverted sensing**
  Taking immediate practical action
  Taking physical exercise
  Looking at the practicalities of the situation.

Inhibitive strategies include

- **Introverted thinking** Analysing and trying to understand exactly what is causing the stress – thinking about it in depth.

- **Introverted sensing** Trying to relax physically, to forget the stress or ignore it by concentrating on other physical activities.

- **Introverted feeling** Trying to cheer yourself up, being with others who have no connection with the stress, being involved in activities that emotionally enable you to forget the stress and feel good.

- **Introverted intuition** Redefining the meaning of a stressful situation, using meditation or imagination to cope.

Stress management

We have looked briefly at the:

- *Causes* of stress,
- *Effects* of stress, and
- *Ways of dealing* with stress.

Stress management brings all these parts together. If you look at what causes you stress, what reactions you have to it and what you do about it, then certain patterns may emerge. However, these patterns may be inappropriate for dealing with the stress because they do not tackle the cause directly.

Two examples follow: David, who is managing stress inappropriately, and John, who is using appropriate stress management.

David's patterns

- *Main cause of stress* Other people, the values of the outside world (extroverted feeling).
- *Effect* Physical tension, illness (introverted sensing).
- *Action* Analysing the stress, theorising about what is causing it (introverted thinking).

This pattern is not going to help David resolve either the cause of the stress or its effect. If other people are causing stress then the best action to deal with it is to tackle the people directly. If this is not possible, then dealing with the effect – physical tension – is the next best step: this would involve physical relaxation. The chosen method – analysis – is singularly inappropriate and yet is used probably because David enjoys thinking and lacks confidence in dealing with people directly.

John's patterns

- *Main causes of stress* John experiences most stress when having to analyse and deal with complex figure work – he runs a small business and the VAT reports, tax and book-keeping tasks are extremely stressful for him (introverted thinking).
- *The effect* Introverted feeling. He feels bad about himself, feels that he is not a 'real' businessman, feels inferior to his accountant and lives in fear of the Inland Revenue.

Now, under normal circumstances John would use extroverted intuition (his favourite function) to try and solve the situation. He would look for more business possibilities, persuade himself that selling to prospective clients was what business was all about, rather than keeping the books accurate. A fairly typical entrepreneur. And of course, stress would just increase, until finally he would have to spend several days sorting out the chaos resulting from his ignoring of the financial details – moaning the whole time.

After being alerted to the pattern of stress management described above, John decided to tackle the cause directly. His first action was to enrol on a three-day 'Finance for small businesses' course. He hated it, but learned all he could from it. He then sat down with his accountant and put into operation a standard set of procedures. He now spends two to three hours every Thursday evening analysing the figures and understanding what they mean. He doesn't particularly enjoy this period of time, but he feels good afterwards at having got an unpleasant task out of the way by approaching it in its own spirit – introverted thinking. Result: stress levels down, a feeling of control, a clearer understanding of the business and more energy available for generating new business – using his favourite function, extroverted intuition, as it should be used.

Go back to your responses to questions 22, 23 and 24 of the questionnaire (pages 21–2). The scores for these questions will indicate whether your response to the stress is appropriate. You may find that, even if they do not match directly, your response is on the right lines. If, for example, you scored 4 for extroverted thinking as a cause of stress, and 2 for extroverted thinking as your way of dealing with it, that is certainly better than not scoring it at all. You need to look at the area for which you *did* score 4 and gauge its appropriateness. If you can, aim to match the function which is causing the stress. If this is not possible, tackle the effect using the appropriate function. Stress, unfortunately, never goes away on its own – and guess who always gets left to tackle it?

6. Don't just work there, *do* something!

The principles of self-adjustment

By now you will have amassed a sizeable battery of self-knowledge: you have a reasonable idea of your preferences, and what you are seen to be good at; you are aware of the implications of these preferences for the way you approach certain tasks and areas of your work; you have had the opportunity to explore your compatibility with your current work situation, and with potential future situations; and you are aware of the danger areas for you regarding stress, and how it can be tackled.

The obvious response to all of this is *so what?*. What good does it do to know you are an engineer with a strong preference for extroverted feeling? Are you supposed to give up your job and become a telesales executive?

We stressed before we started our trip down preference lane that we are not great believers in changing your life for you. We have nothing against an engineer wanting to become a telesales executive; we merely have a passionate professional interest in ensuring that the engineer in question has a clear foundation on which his or her career decision is built.

Self-awareness is a slow-acting medicine, but given ample time and opportunity, it works its way through the system with scrupulous thoroughness. The first thing that should be done with what you have learned from this book is: nothing. The theories we have explored are not particularly complex, but their implications and ramifications can be. Our recommended first step would be to put the book away in a drawer, go back to work and observe yourself. It should not be long before you find yourself approaching a particular task in a way which illuminates a 'light bulb' over your head and causes you to cry out 'I'm using my introverted thinking here!' (or words to that effect).

Once you are aware of the way in which you use your preferences out in the real world, you can return to the material here and confirm how valid you feel your profile to be. Then the real work starts.

What happens next depends really on what you want to achieve. Were you feeling dissatisfied with work without quite knowing why? Had you started a new job, or been promoted, and were unsure exactly how to approach the situation? You need to work on your self-awareness to the point when you are satisfied that you know in which circumstances you use the eight functions well, how you could improve your use of them, and which of them your work particularly demands. If there is a mismatch between what you do well and what your job demands, you move to the Big Decision.

You are at the point when you must decide whether to reframe or retreat. The former requires you to change the way you approach your working life so that your functional preferences can be used in a positive way, and your less preferred functions make up a small part of your life. If that is not possible, you are facing a move – either laterally within the organisation or to somewhere else. In either case you should be asking yourself: Which functions do I want to use more? How could I use them within the kind of work with which I am familiar? Do I need to retrain?

At this point our involvement in your life stops. This book cannot answer the questions which arise once this stage has been reached. But, provided you can perfect the art of observing yourself, and of understanding what you do in which situation and why, you will have gone a long way towards being able to answer these questions yourself.

Before we leave you, we have included some questions to guide your thinking, and to help you build some structure out of the pile of conceptual bricks we have dumped outside your house. Happy building.

Some pertinent questions

- What key issues are you facing at work right now?

- What do you most enjoy about work?

- What gives you most difficulty?

- What are your most preferred functions from your questionnaire results?

- What are your greatest abilities in terms of functions?

- How compatible are these preferences and abilities with your work, your organisation, your team, and the people you work with?

- Given the above, describe your *ideal* work, your ideal company to work for and the sorts of people you would like to work with.

- What actions do you need to take with respect to:
  — your own personal preferences?
  — your skills and knowledge?
  — your environment (present job, organisation)?
  — other people?

- What/who will help you? How might you enhance this help?

- What obstacles will there be? How can you minimise these?

If possible, list targets and actions with times against them.

Readers interested in obtaining the full version of the questionnaire used in this book should write to: Newbase, PO Box 266, Aylesbury HP18 9XT.